New insights into processes of innovation will require a focus on social practice and the dynamic constellations of communities where knowledge lives and circulates. Such a difficult task demands the integration of multiple perspectives. So here is a book that combines several theoretical traditions as well as case studies in different cultures to investigate public policy and cross-organizational innovation as integral part of complex social learning systems.

Etienne Wenger

This book is to be welcomed because it tackles two important needs in knowledge transfer. One is a priority to place discussions about knowledge transfer into an appropriate theoretical framework; the other is a greater understanding of the social context of public–private transactions or collaborations.

Sir Brian Fender
President of the Institute of Knowledge Transfer

Knowledge and Innovation

Although governments across the world are implementing policies to transform the economic base towards high-value products and services, these policies often fail to achieve their intended outcome because the process of knowledge transfer is not very well understood. In this book, Helen Brown presents three case studies from the USA, the UK, and Japan arguing that partnership between public and private sector organizations can take many forms, some of which are very complicated. Vignettes from her case studies explore key attributes of partnerships that are effective in different contexts. Brown argues that 'partnership' should not be prescribed as a panacea for the delivery of complex policy in education, health, and economic regeneration. Instead, policy makers need to adopt a much more subtle and sophisticated concept of multi-agency partnership that acknowledges the time and effort needed to build trust and new shared practices.

Brown takes issue with weak theories of change endemic in some policy, as well as the tacit assumption that knowledge and 'technology transfer' is a logistical problem, that can be resolved by unlocking the flow of knowledge from universities to industry. Instead, the author subscribes to a sociocultural theoretical approach that emphasizes the process of knowledge creation and the significance of consequent changes in the dynamics of human relations. Seen in this light, Brown conceptualizes innovation as collaboration between diverse organizations and individuals, the result of which is organizational learning.

This book will be of great interest to students and researchers interested in policy studies, business and management, and education, as well as policy makers engaged with communities of practice theory.

Helen Brown is the Executive Director of the West Midlands Higher Education Association and a Director of the newly formed Institute of Knowledge Transfer.

RIOT!

Routledge studies in innovation, organization and technology

1 **Innovation in the U.S. Service Sector**
Michael P. Gallaher, Albert N. Link and Jeffrey E. Petrusa

2 **Information and Communications Technologies in Society**
E-living in a digital Europe
Edited by Ben Anderson, Malcolm Brynin and Yoel Raban

3 **The Innovative Bureaucracy**
Bureaucracy in an age of fluidity
Alexander Styhre

4 **Innovations and Institutions**
An institutional perspective on the innovative efforts of banks and insurance companies
Patrick Vermeulen and Jorg Raab

5 **Knowledge and Innovation in Business and Industry**
The importance of using others
Edited by Håkan Håkansson and Alexandra Waluszewski

6 **Knowledge and Innovation**
A comparative study of the USA, the UK, and Japan
Helen Brown

Knowledge and Innovation

A comparative study of the USA, the UK, and Japan

Helen Brown

Routledge
Taylor & Francis Group
LONDON AND NEW YORK

First published 2008
by Routledge
4 Park Square, Milton Park, Abingdon, Oxon OX14 4RN

Simultaneously published in the USA and Canada
by Routledge
605 Third Avenue, New York, NY 10017

Routledge is an imprint of the Taylor & Francis Group, an informa business

First issued in paperback 2012

Typeset in Times by Wearset Ltd, Boldon, Tyne and Wear

British Library Cataloguing in Publication Data
A catalogue record for this book is available from the British Library

Library of Congress Cataloging in Publication Data
A catalog record for this book has been requested

ISBN13: 978-0-415-54158-9 (pbk)
ISBN13: 978-0-415-41663-4 (hbk)
ISBN13: 978-0-203-93935-2 (ebk)

Contents

List of figures vi
List of tables viii
Foreword ix
Acknowledgements xi

1 Introduction and overview 1

2 The theoretical landscape of knowledge transfer 7

3 A natural history of my research methodology 27

4 Reconstructing subject perspectives 44

5 Horizontal expansion of the object by collaboration of experts 54

6 Mediation between activity systems of unequal agency 67

7 The impact of complexity and power dynamics on organizational learning 94

8 Using evaluation as a tool for learning 125

9 Insights from the research findings 147

10 A toolkit to support organizational learning 160

11 Conclusions 175

Glossary 181
Notes 184
References 191
Index 205

Figures

2.1	The transformation of policy during implementation	10
2.2	Positioning HTC policy enactment on the implementation staircase	11
2.3	The structure of a human activity system	18
2.4	Developmental transfer	20
2.5	Depicting structure and contradictions between activity systems	24
2.6	Dynamic tensions in activity	25
3.1	A knowledge-economy policy-borrowing cycle	32
3.2	Tracing the policy-borrowing cycle in case studies	38
4.1	Small-scale textile innovation project	46
4.2	Zero-energy homes: innovation project in California	47
4.3	Hierarchical implementation of HTC policy	48
4.4	Macro-level object of HTC activity	48
4.5	One of the meso-level objects of HTC activity	49
4.6	The micro-level: the object of activity in a project	50
4.7	A depiction of agency at three levels in the enactment of HTC policy	51
5.1	The creation of electronic tools as boundary objects	59
6.1	Communication between powerful and weak activity systems	69
6.2	A depiction of the vertical levels bridged by boundary mediators	70
6.3	Collaborative research into sub-regional skills gaps in ICT	78
6.4	Transformation of a boundary object by its use in new contexts	82
7.1	Constraints on expansive learning between levels	101
7.2	How the NOVUS project links users and developers	109
7.3	Collaboration between physicians and automotive designers	116
7.4	Collaboration between hospitals and the defence industry	119
7.5	How a boundary mediator bridged power difference between levels	121
7.6	The power of an embedded boundary object used by boundary mediators	123
8.1	Guidelines on clustering, 2004	132
9.1	Policy borrowing stimulated by the Silicon Valley phenomenon	148

9.2 Cross-cultural influences on energy policy 149
10.1 Potentially positive effects of difference between organizations 166
10.2 Positive examples from the vignettes 167
10.3 Potentially negative effects of difference between organizations 167
10.4 Negative examples from the vignettes 168
10.5 The role of a facilitator in changing relationships between organizations 168
10.6 Examples of interventions by a facilitator 169
10.7 Tensions between conditions for innovation and partnership 170
10.8 Contradictory factors driving innovation and partnership 171
10.9 The cumulative impact of mini-cycles of expansive learning 171

Tables

3.1	The context and scale of the case studies	31
3.2	Research design including the data collection and analysis plan	36–37
3.3	Emerging forms of expansive learning in the HTCs	39
8.1	Key issues in evaluation	129
10.1	Differences between complex and simple partnerships	164
10.2	Factors critical to successful innovation partnerships	164–165
10.3	Analysis of expansive learning in complex innovation partnerships	172–173
10.4	Examples of mediation by facilitators	174

Foreword

Bridget Somekh

This book engages with some of the most important issues of our day. In a globalized world, regional industries that have provided a stable way of life for several generations can find their economic viability undermined by competition from overseas, and the pattern is repeating all over the world. As the warning signs become clear, governments see it as their business to develop policies for economic renewal and regeneration to avoid disaster for whole populations, but the success of such initiatives is often ephemeral. This book began with Helen Brown's involvement with the community of universities in the West Midlands, instrumental in the high-technology corridors initiative set up by the British Government, and after closure of the Rover car manufacturing company in the West Midlands was narrowly avoided in 1999. Plunged into the day-to-day business of working with universities that had forged new partnerships with local small and medium enterprises (SMEs), she found herself at the heart of a storm of hopes, passions, and resistance. 'Knowledge transfer' was the government's big idea to divert the technological expertise of university academics into the development of new industries and to lead local businesses in developing new products and diversifying their markets in preparation for surviving the next crisis. Yet, the expense and effort of change often seemed an insurmountable barrier, and partnership between the universities and SMEs often involved overcoming cultural barriers between those who saw the world through very different eyes. The Rover Crisis was a precursor of a swathe of government initiatives to promote 'knowledge transfer' which in 2007 is defined by the Institute of Knowledge Transfer as 'The systems and processes by which knowledge, including technology, know-how, expertise and skills are transferred from one party to another, leading to innovative, profitable or economic and social improvements.' In 2005, when I had the privilege of examining the doctoral thesis on which this book is based, Rover had just plunged into another crisis and the end of the car industry in the West Midlands was in sight. Newspapers were full of the good news that strategies for diversifying the industrial base put in place in 1999 made this much less of a crisis than it might otherwise have been. The positive impact of universities as a catalyst for innovation fuelled a greater reliance on 'knowledge transfer' policies.

> The nations that can thrive in a highly competitive global economy will be those that can compete on high technology and intellectual strength – attracting the highest-skilled people and the companies which have the potential to innovate and to turn innovation into commercial opportunity. These are the sources of the new prosperity.
>
> (Introduction to the Science and Innovation Investment Framework 2004–2014)

> We need to do much better at turning knowledge into enterprise and we want universities to see this as their co-mission.
>
> [Gordon Brown, Chancellor of the Exchequer (launch of CMI)]

Helen's meticulous analysis of the development of partnership in the high-technology corridors over the previous five years told a more complex and less euphoric story and made compelling reading. The lens provided by sociocultural theories, particularly *expansive learning theory* and the conceptualization of these willing and unwilling partnerships as different kinds of *communities of practice*, provided fascinating insights into the processes of change. Now, in this book, she has extended and developed this knowledge of the relationship between strategic change policies and (hoped for) socioeconomic renewal, turning her attention to government initiatives in the mature textiles industry and the emerging photonics industry in Japan, and the dot-com companies and sustainable energy innovation in the US. It makes a fascinating study.

This timely book can perhaps be seen as itself a model of the integration of academic and business cultures, since it combines scholarship with powerful, practical insights. The concepts and technical language of expansive learning theory are used (and carefully explained) to bring a rigorous analysis to the apparent confusions of human perceptions and interactions. Knowledge is built up through illustrative vignettes developed from interviews, compared across the different national contexts. Then in Chapter 10, those of us who are concerned with managing change in our own work places are given a rich compendium of good ideas about how change can be facilitated. This is a book which can help governments and managers – in universities and industry – learn how to build partnerships and how to support and sustain change in ways that enrich communities that would otherwise have been broken socially and economically.

Acknowledgements

Thanks are due to all the people who were willing to be interviewed and contribute information, opinions, and insights to this book. In particular, some comments in this study about the role of key actors and agencies in the UK case studies are intended to illuminate the difficulties involved in enacting a complex and ambitious policy. The overwhelming impression was of people doing their very best to work together in challenging contexts to achieve, within a very short time, a huge change.

I am indebted to my critical friends Professor Bridget Somekh, Professor Murray Saunders, Professor Harry Daniels, and Professor Frank Blackler for their comments and advice on the final draft of this book. However, any shortcomings are my own responsibility.

It was a delight to encounter the work of Nuno Corporation, and the insights of Chief Executive Reiko Sudo and Designer Keiji Otani have been invaluable. I must also thank Professor Takaishi and Ms Yoko Ashizawa for their help in translating and interpreting interview transcripts associated with case studies in Japan.

I am also grateful to Vice-Consul, Rachel Lawley of UKTI, for helping me to make contact with key people and agencies in Silicon Valley in the US. In particular, David Aslin and Michael deAngelis were very generous with their time.

The support and bemused encouragement from my family Ed, Emma, and Sarah has spurred me on. Finally, thanks are due to my colleague and friend Sandra Stackhouse for her invaluable support and patience throughout the long gestation of this book.

1 Introduction and overview

1.1 The international policy context

Governments across the world are implementing policies to transform the economic base towards high-value products and services. Innovation policy has been inspired by the success of Silicon Valley based on its proximity to an excellent source of research, principally at Stanford University, and the clustering of high-technology companies in the region. Nevertheless, policies often fail to achieve their intended outcomes because the process of knowledge transfer is very complex and not well understood by policy makers. The outcomes of policy are unpredictable and the implementation of innovation processes involves cultural change in organizations, and it is difficult to manipulate this by policy.

Knowledge economy policy is informed by a weak theory of change. We struggle to grasp how it can be done more effectively and how barriers can be overcome. However, there is a gap between the goals and the implementation of 'technology transfer' policies.

Partnership between private and public sector organizations is presented as a panacea for the delivery of complex policy in education, health, and economic regeneration. Case studies in the UK, the USA, and Japan illustrate the importance of local context in the creation of partnership between diverse organizations to deliver the 'knowledge transfer' or 'innovation'.

Expansive learning theory (ELT) is commonly used in educational settings and only recently has it been adopted in studies of innovation in complex organizations. It is valuable in studies of technology transfer projects because it offers a way of analysing the impact of human dynamics on the process of innovation. It frames the relationships between individual and organizational perspectives through a focus on the use of mediating artefacts and the rules and divisions of labour that shape collaborative outcomes.

Despite the diverse cultural and environmental contexts, both international and regional 'knowledge economy' policies make the assumption that universities can act as a catalyst for change. In a sense, my work focuses on context and brings the analysis full circle with my case study of the financial structures in Silicon Valley and the role of venture capitalists in commercialization. Despite the extensive copying of the Silicon Valley model, the case study investigates some critical success factors that may have been overlooked by those seeking to emulate the success of a unique region.

The case studies illustrate how different types of organizational partnerships affect their capacity for innovation. Organizational partnerships evolve differently in response to unique sets of environmental and cultural contexts. Concepts of partnership are extremely varied, and in different countries, people attach their own specific meaning to the word. It is widely eulogized as an excellent way to bring together the public and private sector to deliver complex policies. Nevertheless, partnerships can vary in complexity. The range and number of partners, the depth of commitment, and the quality of communications affect their sustainability and effectiveness.

The case studies explore the impact of underpinning concepts of knowledge on the way in which a partnership comes to understand knowledge creation, transfer and innovation. Knowledge can be conceptualized as a commodity or as a social process, and these lie at opposite ends of a spectrum of an array of views, which drive policy and practice.

Innovation is often described as three stages: research, development, and innovation. Sometimes the word innovation is used to refer to the culmination of all three stages, from knowledge creation to diffusion of innovation into new technological process services or products.

1.2 My aim and rationale

Policy borrowing is a global phenomenon and remains relatively unquestioned. These contrasting international case studies of innovation focus on 'practices' that are effective in specific contexts. Organizations are composed of people who act and interact and can describe their practices. Regardless of the status or expertise of the respondent, their descriptions of events and practices are informed by prior assumptions, cultural practices, and discourses that 'shape their perspective on reality' (Ball, 1994; Beach, 2003; Lincoln and Guba, 2000; Miles and Huberman, 1994; Ozga, 1999; Silverman, 2001; Yin, 2003).

Practices and policy can form a virtuous circle but only if the evaluation of practice creates evidence to inform the development of more successful ways of doing things.[1] In other words, the true value of evaluation lies in the opportunities to *learn* during the implementation process and to adjust the trajectory of policy accordingly. The case studies are analysed to try to answer five research questions:

1 Why is it that some approaches seem to have evolved in a particular geographic and social context?
2 What can be learnt from such exemplars and what if anything can be transplanted or adopted in a different context?
3 What are the benefits and limitations of using ELT to analyse the development of multi-agency learning?

4 How can boundary objects and mediators facilitate innovation?
5 What insights from ELT and from the analysis of collaboration in these international case studies can be applied to similar policy initiatives?[2]

In the past ten years, my interest and professional involvement in regional policy development has grown. For much of the last century, the West Midlands region of England was the heart of UK manufacturing industry and car production. During the past twenty years, increasing global competition had a massive impact on the competitiveness of UK manufacturing companies. I was fascinated by the way in which a regional innovation policy evolved in response to the crisis at the Rover Car Manufacturing plant in 1999.

I took the opportunity to study the 'high-technology corridor' (HTC) policy as an example of an attempt to enhance the impact of higher education institutions on the economic regeneration of an English region. The policy describes a mechanism for universities to contribute to the so-called 'knowledge economy'. From the perspective of universities, this belongs to a type of activity known as the 'third mission',[3] described variously as 'knowledge transfer' or 'technology transfer'. HTCs are geographically pre-determined sites of interaction between sources of research and commercial exploitation that aim to drive a sub-regional knowledge economy as described in the West Midlands Regional Innovation Strategy.

My experience led me to question theories of policy development, which describe a smooth and rational process. I wanted to see if national innovation policies were implemented more easily elsewhere. Two US case studies illustrate the development of partnerships between academia and commerce to drive innovation. The first of these explores partnerships founded to commercialize sustainable energy technologies. The second revisits the 'Silicon Valley phenomenon'. Silicon Valley is an exemplar of research-led innovation, but the name also conjures a powerful metaphor for successful economic regeneration driven by an alliance between universities and high-technology companies. My case study concentrates on the environment in which innovation can flourish in terms of financial services and attitudes in Silicon Valley.

Two Japanese case studies describe innovation practices in a mature industry sector and in an emergent sector. The first explores long-standing partnerships between an innovative design company and various small- and medium-sized manufacturers (SMMs). The Japanese textile industry has an international reputation for the creation of new materials, products, and processes. My second case study describes the Government and regional innovation policies that support collaborations between universities and a cluster of photonics companies.

ELT has gained currency in postmodern studies of organizations. It has been used as a framework to understand how agencies connect in an organic way, and in some cases, research has focused on innovation within large complex organizations. Unlike predominant theories of innovation, which underestimate the extent to which the process of 'knowledge transfer' relies upon the quality of relationships between people, it offers a heuristic framework in which this is of central importance. My case studies describe opportunities for expansive learning by individuals and organizations working in partnership in terms of variation in communication, trust, and clarity of purpose.

My work builds on several well-referenced and recent studies that have used ELT (a variation of activity theory) to inform their understanding of innovation and organizational change in which previously disparate work teams have devised ways of collaborating (Blackler, 1995, 2000; Blackler *et al.*, 1999, 2000; Engeström, 1999, 2000, 2001, 2004; Engeström *et al.*, 1999; Kerosuo and Engeström, 2003; Ludvigsen *et al.*, 2003; Toiviainen, 2003).

Although the theoretical concepts are developed in the context of a study of innovation, the research methods and analysis may be of interest to researchers and professionals in the fields of education, social work, management, and healthcare. To enable a wide range of readers to dip into particular chapters of this book, there is some reiteration of key theoretical points.

1.3 The structure of this book

This book addresses a significant gap in theory. Currently, 'expansive learning' is conceptualized as collaboration, but this does not adequately describe learning between communities where agency and power vary. However, these case studies reveal the impact of power as a structural constraint to multi-agency learning.

Chapter 1 introduces the cases and situates them in their local and national policy context. It discusses the international policy context for the 'knowledge economy' and the assumptions made in global and local policy.

Chapter 2 outlines the theoretical landscape of innovation and discuses policy implementation by partnerships. There is a tacit assumption that knowledge and 'technology transfer' is a logistical problem, to be resolved by unblocking the flow of knowledge from universities to industry. By contrast, sociocultural theory emphasizes the process of knowledge creation and the significance of consequent changes in the dynamics of human relationships. Following this line of argument leads towards a conception of innovation as collaboration and organizational learning.

Chapter 3 outlines my research design and methodology, linking this to the heuristic framework of ELT. My case study approach recognizes that policy is not necessarily implemented as a linear process and innovators may not even describe what they do as 'knowledge transfer'.[4]

Chapter 4 draws on the empirical data to reconstruct various 'subject perspectives' in relation to changing conceptions of the 'object of activity'. Variations in perspectives show how key individuals and networks of 'partner' organizations are influenced by various competing 'objects of activity' that emerge during the policy implementation process.

In Chapter 5, I discuss the capacity of boundary objects to act as bridges between experts from different 'communities of practice'. In such instances, this leads to the horizontal expansion of the 'object of activity'. I present the data as vignettes that sketch the different ways that material artefacts and 'conceptual models' act as boundary objects or tools to mediate learning across different activity systems. Several vignettes illustrate how boundary objects mediate between the perspectives of users and developers of new technologies in innovation projects.[5]

In Chapter 6, I focus on the impact of complexity and power dynamics on the capacity for organizational learning. Examples from the empirical data illustrate contradictory sets of rules operating at different levels of policy enactment. Activity at macro-, meso-, and micro-implementation levels is considered in terms of characteristically different relationships of power and politics. Bureaucracy is visualized as a powerful kind of inertia that can confound the espoused policy aim to bring the public and private sector closer together. Vignettes or pen portraits focus on mechanisms, which connect the 'perspectives' of individuals and organizations and expand the object of activity towards a new and shared perspective. I use diagrams to recreate the object of activity at different levels and show how special people can act as mediators and enhance the process of technology transfer by connecting a vertical hierarchy of poorly linked activity systems.

Chapter 7 describes the combined impact of boundary mediators and boundary objects. Together, they act as bridges between horizontally and vertically arrayed activity systems and are more effective in mediation. I show how discrete 'knowledge transfer projects' have the capacity to contribute to fuller expansive cycles of collaboration and learning. I draw on key insights from the data analysis to encapsulate how particular incidents and practices were 'experienced' by different individuals and groups. A thematic discussion of dilemmas and unresolved tensions in activity highlight various policy implementation problems.

Chapter 8 focuses on the use of evaluation as a tool for organizational reflection and learning. I discuss the political sensitivities of evaluations and the need to encourage practitioners to share in the evaluation process. I also discuss the risk of distortion by policy makers with a vested interest in justifying the investment of public funds and intent on a victory narrative.

Chapter 9 summarizes the key points and issues from my research findings. Chapter 10 proposes a toolkit for project managers to assess the weaknesses and strengths in collaborations between different organizations. The case studies show that the development of robust and sustainable organizational partnerships cannot be taken for granted. The toolkit can be used to guide the investment of future efforts in partnership building.

Chapter 11 reflects on my conclusions about the current gaps in ELT, which make it difficult to differentiate the quality of collaborative practices. I discuss the value of integrating some insights from complexity theory and communities of practice theory (COPT). I summarize my theoretical insights including a discussion of the circumstances that inspire individuals and organizations with a shared passion and commitment. I also summarize some practical proposals for using configuration devices to improve the sustainability and durability of collaborative practices. In identifying effective strategies and potential hazards, I hope that all the people who contributed to this research by sharing their experiences may help similar policy initiatives to be implemented more smoothly and with less anxiety.

Turning now to Chapter 2, I set out the theoretical underpinning of 'knowledge economy' policies and the assumptions about 'organizational partnerships' as a popular policy delivery mechanism.

2 The theoretical landscape of knowledge transfer

Knowledge transfer policies can be considered as experiments in social engineering aimed at economic regeneration through innovation. This chapter highlights contradictions and inconsistencies in the policy assumptions. Governments have approached the challenge of 'innovation' and 'knowledge transfer' with a loose assemblage of theoretical concepts rather than an integrated and coherent approach to policy implementation. In other words, much technology transfer policy has been informed by a rather weak theory of change.

2.1 Theoretical portraits of the innovation process

Theories can act as lenses to illuminate different aspects of complex processes. In many ways, the lenses are complementary rather than mutually exclusive, and where the conceptual frameworks converge, there are some interesting ways in which the different perspectives add value to one another to give a more complete picture of the landscape of innovation.

Social theories can be seen as sensitizing frameworks for empirical research in the social sciences. In this sense, they can open out 'ways of seeing and analysing social phenomena' and, at the same time, mould and change our self-understanding (Reckwitz, 2002: p. 257).

Many influential theories of knowledge transfer underestimate the impact of the quality of relationships between people and the need to develop and sustain communication. Consequently, I have designed my case studies to capture some important aspects of social and cultural interaction, which are often underplayed in comparison with the economic and technological characteristic of innovation. From a sociocultural perspective, technology does not exist independently of the people who create it. Therefore, it places a much stronger emphasis on the impact of human dynamics.

These case studies have allowed me to explore the capacity of 'expansive learning theory' (ELT) to provide a coherent and consistent account of the emergence of collaborative practices. In this sense, the theory is also treated as an object of study. It is used to frame the tensions and dilemmas arising within and between activity systems.

Innovation involves changing and breaking familiar structures and habitual routines. In ELT, innovation occurs when traditional ways of doing things prove to be inadequate to the challenges of new situations. This results in tensions that drive the development of new practices. The subsequent tensions between the old and new ways of doing things then drive further change and innovation.

Sociocultural theory emphasizes the process of knowledge creation and leads towards a conception of innovation as collaboration between diverse organizations and individuals resulting in organizational learning. From the perspective of ELT, the expansive learning cycles are the process by which the object of activity can be expanded through the collaboration of two or more different communities.

The ontology of the enquiry assumes that prior understanding and assumptions shape reality for individuals and the stakeholder groups to which they belong. This orientation is consistent with finding out how an innovation project is 'experienced' by individuals representing major stakeholder groups. Social constructivism is often termed 'critical realism' or 'fallible realism', and the enquiry adopts this approach. It aims to 'account for events rather than simply document their sequence … and to provide a causal description of the forces at work' (Miles and Huberman, 1994: p. 4).

The research addresses the ontological, epistemological, and methodological problems of 'seeing organizations' by adopting the notion of practice as the trajectory of organizational learning. Practices can be defined as 'sets of acts and interactions' involving language and objects (Giddens, 1993; Saunders, 1995; Yanow, 2000). The concept of practice enables organizations, such as complex multi-agency partnerships, to be perceived in terms of relationships between concrete practices.

It is interesting to compare and contrast four models of innovation, which emphasize different aspects. Each of these perspectives designates a different aspect or entity of innovation as the target and locus of the process. In describing the innovation process, Van de Ven's chaos perspective focuses on the emerging product, Nonaka's knowledge creation perspective focuses on emerging knowledge, Leonard-Barton's core capability perspective considers emerging capability, whereas Engeström's learning perspective concentrates on the object of activity. Essentially, each theory uses different words to describe the process of innovation in four stages. The first stage involves the formulation of the problem or idea. The second stage involves modelling a publicly transmittable representation of this. In the third stage, the model is tested in simulation or real life. Finally, in the fourth stage, the model is consolidated and stabilized in the acquisition of new practices.

Despite the discrepancies between these theories, they all support the view that the development of new ideas into knowledge, products, and associated practices is a time-consuming, capricious process that surpasses the reach and ability of any individual. They emphasize the emergent, contingent, and ambiguous nature of innovation and agree that a 'sense of progress' distinguishes innovation from destructive chaos (Engeström, 2002).

2.2 Theories of policy change and implementation

Theories represent the stories that people tell about how problems arise and how they can be solved. Stories can arise from stereotypes, myths, or research knowledge, and they are potent forces in policy discussion. Policies that seem to violate the assumptions of prevailing stories (dominant discourses) will receive little support (Weiss, 1995). It is also important to recognize that how we analyse data or phenomena affects what we find out. In other words, our methods affect the patterns we find as well as our interpretation of their meaning. Policy change is likely to be most successful when informed by robust theory and evidence. However, complex innovations are difficult to implement especially when the innovation is undertaken by a multi-agency partnership (Fullan, 1999).

National governments have devolved much of the responsibility for innovation policies to regions. Consequently, it is possible to compare the implementation of 'local experiments' to transform an industrial and manufacturing region into a knowledge economy. Different types of policies are affected by different contextual conditions and carry with them different possibilities for implementation. Studies of implementation processes must therefore take account of specific local contexts in which policies are implemented (Bredgaard *et al.*, 2003: p. 1) . The case studies were analysed to explore gaps between theory and practice in policy development and implementation.

Change happens at local levels during the implementation process. Innovation policy is not unusual in embracing a vast range of sites of action and discourse. It is normal for policy to pass through many long and elaborate implementation chains (Halpin and Troyna, 1994). Haphazard approaches to change are likely to be frustrating, partly because they do not prepare innovators for the experience of the implementation gap (Trowler *et al.*, 2002). Implementation can be described as 'what really happens in practice' (Fullan, 1993). The process of change is inherently volatile and dynamic because as each change is introduced, it brings about further unpredictable changes.

The change process can be smoother if policy makers think carefully about what they are trying to achieve and appreciate the limits of top-down efforts to create change. It involves persuading people to abandon familiar, comfortable routines, and the process of creating new practices can be slow, incremental, and unpredictable (Trowler *et al.*, 2002). Change is constructed by people in specific contexts, and it can be difficult to '*unfreeze* practices' in order to make way for new ones (Trowler *et al.*, 2002). Ambitious and significant changes take much longer to embed than anticipate.

Although policies are attempts to create a new reality, a policy vision is often distinctly different from its effect, and there is a risk in being too rigid in prescribing for change (Fullan, 1993; Ozga, 1999: pp. 94–95). The process of policy evolution and practical change is analogous to the notion of sense making. The implementation process needs to allow flexibility for aspects of the policy to be rejected or modified by experience. When the impact of a new practice is understood, it is important to be able to adjust the practice based on an evaluation of its actual effect compared with the desired effect. This flexibility allows 'practitioners' to contribute to a shared process of sense making and helps to ensure that the innovation is sustained and assimilated.

Policy aims are not transmitted and received in a one-off process because negotiations and reconceptualization occur between agents and at different levels despite the unequal power relationships (Reynolds and Saunders, 1987: p. 213). The implementation staircase model suggests that policy enactment is capricious and is affected by people and factors at various positions and levels in the process. It models how policy is 'transformed', as the policy is enacted by local practitioners. It focuses upon the subtle and often invisible ways policy 'changes' to make it 'workable' in a specific local context. It shows how feedback and reflection on practice affects the development and the trajectory of the policy as it takes account of various levels including those above and below the key agents.

In practice, policy development and implementation is often a turbulent and conflictual process. Figure 2.1 illustrates a scenario in which the trajectory of innovation policy can be informed by feedback during the process of implementation.

The implementation staircase articulates a problem that is often ignored, namely

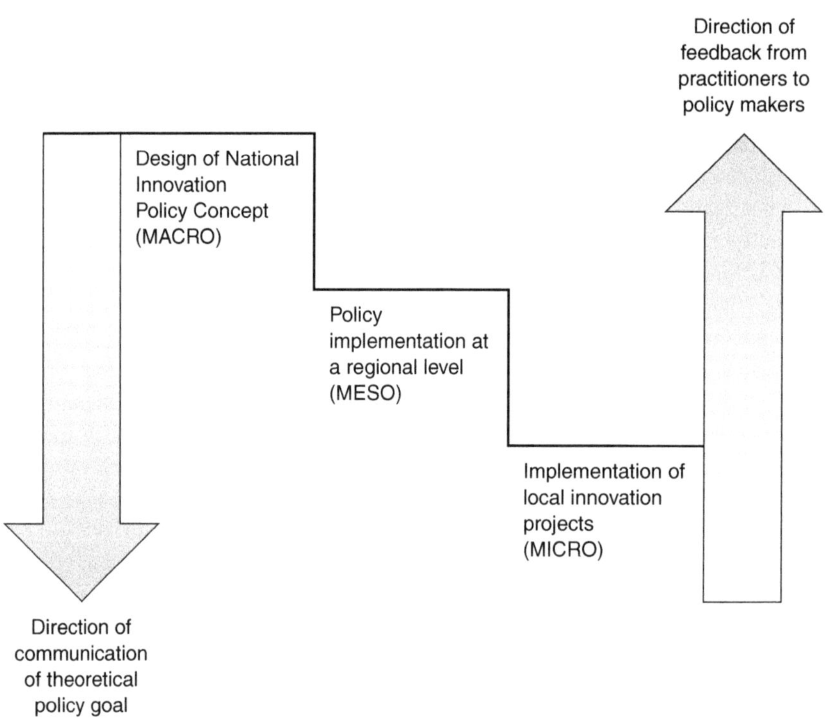

Figure 2.1 The transformation of policy during implementation.

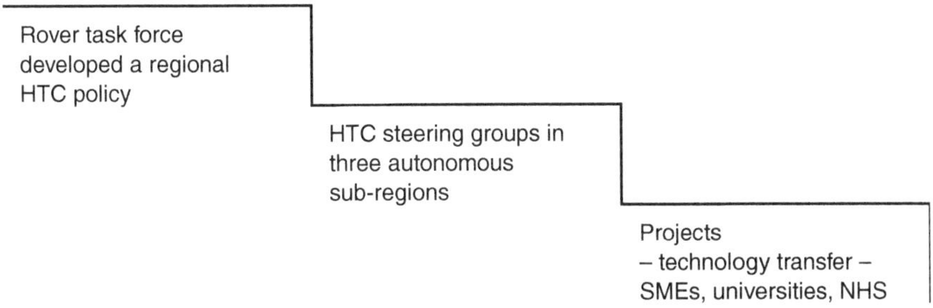

Figure 2.2 Positioning HTC policy enactment on the implementation staircase.

the dissociation of policy discourse from practice. It recognizes the process of negotiation between practitioners and policy makers. It describes the possibility of movement upwards and downwards in the hierarchy of decision-makers at strategic and operational levels. It was useful to consider the relationships between hierarchical levels in the complex UK case studies, and it was possible to adapt the implementation staircase model to depict three predominant levels of policy enactment as shown in Figure 2.2 adapted from Reynolds and Saunders (1987).

A similar understanding of the adaptation of policy as it passes down and up the implementation staircase is adopted in an analytical concept called policy refraction. It depicts the way policy is re-conceptualized and adapted at different levels and sites and acknowledges the key role played by practitioners in policy development (Higham *et al.*, 2002: p. 10). This flexibility and openness to policy adaptation by people closest to the ultimate beneficiaries of policy is important.

2.3 Policy assumptions about triple helix partnerships

Partnership between public and private sector organizations is presented as a panacea for the delivery of complex policy in education, health, and economic regeneration.

> The influence of Triple Helix Theory can be detected in the majority of national innovation policies. A triple helix regime typically begins as university, industry and government enter into a reciprocal relationship in which each attempts to enhance the performance of the other.
>
> (Viale and Etzkowitz, 2005: p. 6)

It can also be traced that, in the strategic plans, many leading universities have developed for their Third Mission work (e.g. Melbourne Institute of Technology's Mission Statement). This is closely allied to particular interpretations of complexity theory as it migrated from its original home in mathematics and science to management schools and was applied to studies of organizational behaviour in companies.

The effectiveness of innovation as a driver of economic growth in the USA has led to the phenomenon of policy borrowing. There is a tacit assumption that knowledge and 'technology transfer' is a logistical problem, to be resolved by unblocking the flow of knowledge from universities to industry. However, 'knowledge transfer policies' are littered with contradictions and inconsistencies. There are few examples of theoretically consistent and coherent approach to policy implementation. Governments across the world have been influenced by triple helix theory, which draws on some aspects of complexity theory in its analysis and prescription for national innovation policy.[1] It focuses on the changing nature of knowledge-based innovation systems in the light of the dynamic interconnections between the university, industry, and government.

> Most such initiatives take place at the regional level to address problems in industrial clusters, gaps in academic development and lack of governing authority. The first step usually involves collaboration among universities, firms and governments in a project to enhance a local cluster or create a technopole.
>
> (Viale and Etzkowitz, 2005: p. 6)

Industries have to assess in what way and to what extent they decide to internalize R&D functions. Triple helix theory has had a global influence on the way in which governments intervene to catalyse innovation. Universities position themselves in markets, both regionally and globally.

> In the Triple Helix model constructed advantages have been conceptualized as the surplus value of an overlay of relations among the three components of a knowledge-based economy: (1) the knowledge-producing sector (science), (2) the market, and (3) governments. Those places with research universities witness a growing demand for knowledge transfer to industry and, through government, to society.
>
> (Cooke and Leydesdorff, 2006)

The relationships are likely to be affected by turbulent interaction of factors such as the marketability of the technological innovation in a global context and the geographic variations arising from specific local and regional conditions (Leydesdorff, 2005a, 2005b, 2006; Etzkowitz and Leydersdorff, 2007; Leydesdorff *et al.*, 2006).

Leydesdorff acknowledges the value of conceptual linkages between different theories of innovation, and he adopts the concept of translation from actor-network theory (ANT). He also refers to fuzzy boundaries and dynamic change and instability, and there is some resonance with the notion of creative destruction in ELT. Stability exists at the same time as instability, and something new emerges from what existed. This paradox is a significant feature of both complexity theory and ELT (Stacey, 1996, 2003). However, 'creative destructions' reach across previously established boundaries (Leydesdorff, 2005b).

> The entrepreneurial university makes science but also technology and increasingly innovation. In the twentieth century there is a change of the process of innovation with the emergence of corporate, university, and government sponsored R & D, called by Mowery and Rosenberg (1998) the 'institutionalisation of innovation'.
>
> (Viale and Etzkowitz, 2005: p. 5)

Governments in the EU and Japan have intervened with public funding to support clusters of companies around centres of research excellence. However, in many instances, the successful recipe is copied only partially, and the uniquely market-driven innovation around Silicon Valley and Boston is virtually impossible to replicate elsewhere. 'New knowledge increasingly appears in "polyvalent" forms, with theoretical, practical and interdisciplinary implications forming a common centre of gravity, the DNA of the triple helix' (Viale and Etzkowitz, 2005: p. 5). For instance, the US case study suggests that research initiated by a private sector company, working with a university to develop a specific technology from fundamental research is a rapid route to commercialization. Many firms find they must collaborate with other firms to share the risk and cost of competing globally. This has been popularized in cluster theory developed at Harvard University (Porter, 1998).

> New business activities through moderate 'collaboration' are a phenomenon not confined to Japan. In an industrial cluster in northern Italy, which boasts one of the highest growth rates in Europe, enterprises are actively engaged in modest collaboration through a collaborative structure called 'Piattaforma'. Activities are undertaken in a variety of fields, including apparel, automobiles and machinery.
>
> (JSBRI, 2005)

The challenge to collaborating firms is in actually realizing those economies of scope and scale through the integration of their efforts. This book addresses an important aspect of that integration problem, namely the transfer of technologies or knowledge from one site to another. The dilemma facing any manager in a cooperative venture is first, whether the partner has something of benefit to offer, and second, whether that benefit can be appropriated back to the home firm (Rebentisch, 1994). The economic development policy of the Japanese government identifies social capital as the basis of collaborative ventures and takes active steps to encourage triple helix collaborations.

> These developments are taking place primarily in highly competitive industrial clusters in Europe and elsewhere, such as Silicon Valley in the U.S. In business, research and industrial clusters, informal ties of trust (cooperative relations) form between enterprises, universities and other entities.
>
> (JSBRI, 2005)

However, EU innovation policy has to be interpreted in more than twenty-five nation states. The complexity of setting down clear rules is bound to be more challenging than in the US and Japanese national innovation systems. The impact of the bureaucracy associated with this increases the risk that the EU system of public funding hinders the very process it sets out to catalyse.

In recent years, global changes in supply chains and networks led to reversal in the patterns of import and export of manufacturing goods and high-technology products and services. This has been challenging for many developed economies.

> The last few decades have witnessed a dramatic shift in the manner in which business is conducted around the world. Firms have shifted away from a hierarchical, one-dimensional supply chain entity to a fragmented network in favour of strategic partnerships with external entities. This global phenomenon causes ripple effects throughout the old supply network.
>
> (Bitran *et al.*, 2006)

Japan has experienced difficulties in retaining the Keiretsu – the supply chain in small- and medium-sized manufacturing companies (SMMs). New more flexible structures have evolved to replace this method of obtaining a competitive advantage.

2.4 The nature of knowledge: discourses in sociocultural theory

The quest to transform traditional manufacturing economies into 'knowledge economies' has become an international policy imperative. The financial risk of failing to develop ambitious policies contributes to a tendency to take urgent policy action without fully considering the nature of knowledge itself.

A discourse can embody the position from which an individual or group speaks. In daily life, discourses compete for dominance, and strong discourses are often backed by money and power. Discourses 'work and compete' within social practice. Policy documents can be described as *discursive objects* because they are formed by social practices such as making and obeying rules. Discourses have a social basis and can embody the power relationships between individuals and institutions that encourage or constrain them.

> Discourses are partial and positioned, and social differences are manifest in the diversity of discourse within particular social practices. Evaluating discourses means setting them against shifting understandings of what material possibilities there are in the practical domain concerned – discourses are not evaluated against 'absolute truth' but in terms of 'epistemic gain'.
>
> (Fairclough, 1999: p. 74)

Postmodernism[2] recognizes the importance of 'local' discourses. Different discourses are associated with specific perspectives or mindsets which reflect the 'positioning and interests' of the organization or community to whom an individual owes allegiance. Foucault stressed that local discourses are affected by global and national discourses, and so policy documents can be used to enrol and govern individuals and groups (Foucault, 1972).[3] 'Foucault proposes that discourse is not to be treated as a mere "document" of the metal qualities "behind it", but as a sequence of external events in which symbolic structures (formative rules) are manifested' (Reckwitz, 2002: p. 248). Unlike theories that visualize knowledge as a product, the social constructivist discourse in sociology represents knowledge as a process. In this sense, both an innovation project and an innovation policy can be considered in terms of the social relations it can draw people into and the individuals and organizations that stand to gain or lose.

Within sociology, there has been much recent interest in theories of practice, and consequently, organizational routines are of significant interest. From a sociocultural perspective, *situated practice theories* focus on the dynamics of everyday life and the choreography of relationships between peoples and agencies. Taking a different approach, *theories of social structure* focus on institutions, rules, and norms. Their interest lies in the analysis of social systems, discourses, and history, and in the most extreme cases, they deny that individuals are 'knowledgeable' and have *agency* in the sense of the power to act independently of social structures and group norms.

Cultural theories, including practice theory, seek explanations for action in symbolic structures of meaning and language, discourses, and conceptual tools or technologies. Although practice theory is an example of cultural theory, not all cultural theories are practice theories. Models of practice draw on the ideas of Wittgenstein, Giddens, Latour, Foucault, and Bourdieu among others and are interested in the 'everyday' and 'life world' (Reckwitz, 2002: p. 244).

Practice theory regards agents as carriers of bodily routines of behaviour and mental routines of understanding and knowing embedded in the use of objects. Carrying out a practice very often means using particular things in a certain way. A specific social practice contains specific forms of knowledge. This is more complex than 'knowing that'. It embraces ways of understanding and knowing ways of wanting and of feeling that are linked to each other within a practice (Reckwitz, 2002: p. 253).

A practice is a routinized type of behaviour, which consists of several elements, interconnected to one another, including know-how, states of emotion, and motivational knowledge. A practice, such as a way of cooking for instance, cannot be reduced to any one single element. For example, a professional chef carries out social practices, and as such, they are neither autonomous nor constrained to conform to norms (Reckwitz, 2002: p. 256). The relationship between practices encompasses relationships between individuals and communities mediated by artefacts, social processes, and structures. In other words, the practices adopted by a chef have a history and a cultural context and are rooted in a specific community of practice.

Weick's theory of 'sense making' shares a phenomenological focus and speaks of social activity as grounded in 'physically situated routines'. The process of 'sense making' happens when 'people make retrospective sense of the situations in which they find themselves and it is this process that shapes organizational structure and behaviour' (Weick, 1995).[4] In Giddens' structuration theory, and Lave's situated learning theory, organizational routines are emergent and enacted (Giddens, 1993; Lave, 1993; Weick, 1995). Giddens introduced the notion of the 'trajectory of the self' in which people reflexively reconstruct their self-narratives, to create coherent relationships between the various sequential and disjointed influences on their lives. He suggests that individual identity is an emergent self-image, which is crafted and mediated by changing cultural factors of various social groups to which the individual belongs (Giddens, 1993). The problem of distinguishing between individual and collective identity is discussed more fully in Chapter 4 in relation to the empirical data.

ELT is based on cultural and historical activity theory (CHAT), and it has been used in studies of organizational change to emphasize the interplay between individual and collective action. Organizational culture can be seen in terms of processes created by individuals and groups whose activity and motivations are shifting and ambiguous. From the perspective of ELT, it is possible for both individuals and organizations to learn. This is consistent with the view that knowledge is constructed by a complex interaction between individuals and communities. The case studies conceptualize innovation projects as networks of activity systems. Each 'activity network' can be visualized as a meta-activity system comprising relationships between sub-activity systems.

Knowledge cannot therefore be transferred because it is not a stable material but deeply contextual. It is located in time and space and is constantly constructed and recreated by people. A recent report on innovation commissioned by the Organization for Economic Co-operation and Development (OECD) describes the need to transport the knowledge from the researchers in universities to industry where it can be applied in commercial settings (Goddard and Chatterton, 1999). Although it differentiates between codifiable (know-what) knowledge and tacit (know-how) knowledge, there is no adequate explanation of how to transport tacit knowledge.

Much research has shown that it is difficult to gain access to tacit knowledge (Blackler, 1995; Fox, 2000; Gherardi, 2000; Robertson *et al.*, 2003; Yanow, 2000). For instance, even though 'organizational routines' can be carriers of tacit knowledge, it is not easy to transplant routines or 'know-how' from one work environment to another (Giddens, 1993). New practices evolve continuously and are in tension with old practices.[5]

This concept is close to the knowledge management cycles described in the recent studies of creativity, which bring together aspects of organizational learning theory with the theoretical aspects of complex adaptive systems (CASs). The earlier generations of knowledge management theory (KMT) tended to focus on the form of knowledge and on rules for action, which could be imitated. The next generation of theory moved towards a conception of knowledge as a process and focus on know-how as well as know-what. Senge, Stacey, and Holland developed a model of CASs to focus on education and innovation. 'For me knowledge is not a tool; knowledge is meaning. We might use tools like books or databases or whatever in our interaction with each other, but for me knowledge emerges as meaning in the interaction between us' (Stacey, 2002: pp. 10–11). Sociocultural theory has shifted the focus of discussions of 'knowledge management' away from issues of information and technology towards those of human capabilities (Wenger, 1998, 2000). During the last century, organizations have moved from a focus on mass production towards mass innovation. Continuous processes of innovation and mass customization are moving closer to experimentation (Ludvigsen *et al.*, 2003). However, it is difficult to articulate 'the process of innovation' because it is locally generated as 'professional knowledge'. This tends to be not only local but also tacit (Colville *et al.*, 1999). ELT proposes a conceptual framework for knowledge creation that embraces the concerns of both *situated practice* and *social structure* orientations. ELT describes the process of collaboration as organizational learning and co-configuration. In this sense, 'knowledge' can be treated as a shared social construction by organizations working together.

2.5 Innovation as collaboration and expansive learning

In Scandinavia, ELT is becoming a common paradigm for analysing complexity in the workplace and claims to bridge the traditional divide between practice and theory (Blackler *et al.*, 1999). It conceptualizes innovation as a process of collaboration instead of the unproblematic movement of information from one location to another. The term 'co-configuration' is used to describe the shared co-construction of new practices (Tuomi-Grohn and Engeström, 2003). ELT has been used to conceptualize the ways in which experts can work together by applying complementary expertise to a new problem that they could not have tackled independently.

Recent studies have depicted this as interactions between equally important but distinctive activity systems. It is described as a process of *horizontal learning* in which good communication is essential to the negotiation of expanded practices (Hasu, 2001; Kerosuo and Engeström, 2003; Ludvigsen *et al.*, 2003; Toiviainen, 2003; Tuomi-Grohn, 2003). Many of these studies have been conducted in structured and facilitated environments in which researchers have intervened to influence the dynamics and trajectory of change (Engeström, 1999, 2001, 2004; Engeström *et al.*, 1999; Hasu, 2001; Miettinen and Hasu, 2004; Toiviainen, 2003; Tuomi-Grohn and Engeström, 2003).[6]

Within the last three years, 'complex organizations' have been the subject of study by researchers associated with the Change Laboratory at the University of Helsinki (Collins *et al.*, 2004; Engeström, 2004; Hasu, 2001; Ludvigsen *et al.*, 2003; Miettinen and Hasu, 2004; Toiviainen, 2003). Researchers traced and visualized disturbances and breakdowns in everyday work practices in order to reformulate or expand work practices. These studies were conducted in professionally facilitated environments in which the researchers made deliberate attempts to develop collective work practices. Researchers intervened to help people to visualize tacit knowledge and emerging practices to support and accelerate a smooth transition towards the development of expansive learning (Engeström, 2004; Kerosuo and Engeström, 2003). Engeström's recent studies of innovation in California involved analysis of videotapes of team meetings and interactions at work, which allowed: 'very detailed data driven analysis of the discursive processes, practical actions and mediating artefacts that are employed in the step-by-step production of an innovative solution or idea' (Engeström, 1999: p. 377). In developing ELT, Engeström applied CHAT to new contexts, some distance from the original focus of Russian psychologists on concept development in children. During the 1920s and 1930s, Vygotsky's pioneering work in activity theory was concerned with individual learning and development. Activity theory was influenced by insights of Marx and Hegel that tools could shape the progress of cultural evolution and could help to create more advanced ways of doing things.[7] In the concept of *practical consciousness*, Marx makes an important connection between action and thinking of relevance to innovation. '[A] division of labour between thinking and working risks creating a social production of knowledge that grossly misrepresents reality' (Hellström, 2004: p. 641). At the heart of activity theory is an interest in the capacity of tools to *mediate* in the relationship between the 'ideal' and the 'concrete'. In other words, 'what is imagined' can be realized with the help of a tool. Leont'ev designed a triangular diagram to show how a tool mediated the relationship between the subject and object of activity.[8] Later, Vygosky's work in the field of education, focused on the use of language as a mediating tool in human activity. Engeström adopted and changed the model and arrayed collective elements within an activity system in which the mediating artefact could be a physical object, mental model, or symbol system, such as language. Engeström moved beyond the original focus on individual learning and mediation of human

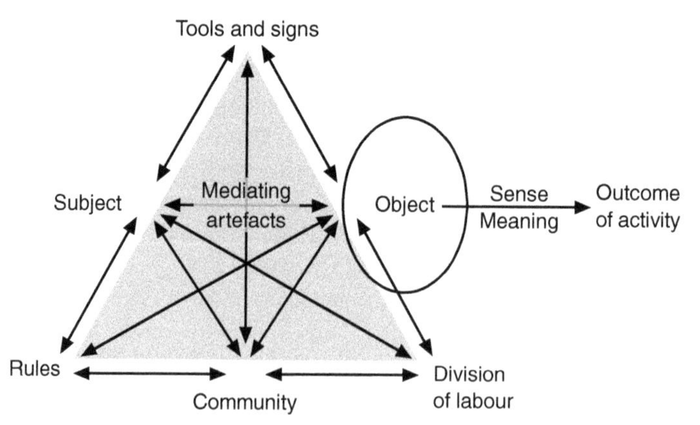

Figure 2.3 The structure of a human activity system (Engeström, 1987: p. 78).

consciousness to re-orientate CHAT towards studies of organizational learning. 'Activities are social practices oriented towards objects, motivated by a human need' (Fichtner, 1999: p. 380). Figure 2.3 depicts the structure of an activity system and situates individual action within a social context. It shows how tools such as concepts and technologies mediate the interaction between the individual and the emerging object. At the same time, a division of labour mediates the interaction between the community and the emergent object of activity leading to an eventual outcome.[9]

ELT encapsulates the idea that, as a result of people working and thinking together, organizations have the capacity to change. Organizations have a developmental trajectory in so much as future activities are created by past activities. It suggests that conflict is creative because new and better ideas push the developmental trajectory towards a more comprehensive understanding. Crossing the boundaries between the surrounding worlds or activity systems brings various subjects and objects into creative tension, and the tension between old and new practices drives the innovation process (Hellström, 2004: pp. 639–640). 'Innovation is not just the conception of a new idea, nor the invention of a new device, nor the development of a new market. The process is all those things acting together in an integrated fashion' (Hellström, 2004: p. 634). Innovation involves both mental and physical activity; it concerns a problem and its solution and involves the creation of something new. The notion of innovation as a creative and conflictual process is influenced by a theory of dialectical action introduced by Hegel. Innovation from the point of view of Hegel's dialectics implies that progress does not happen in a smooth and orderly way (Hellström, 2004: pp. 639–640). A new idea is created in opposition to existing ideas and practices, in which conflict and power struggles play a central role. This contrasts with notions of innovation as 'a linear model of continuous improvement' popularized in the Japanese concept of 'Kaisen'. The conception of innovation as a disembodied 'thing' detached from human activity is misguided. 'A result of the reification of innovation and the emphasis on innovation management rather than innovation dynamics product developers become 'cogs' in the machinery of the innovation system' (Hellström, 2004: p. 643). As a result, relations and selves become more 'thing like' and the social relations of innovation more manipulative (Hellström, 2004: p. 643). In descriptions of activity systems, the 'subject' sometimes refers to an individual acting as a representative of a small group and sometimes the term 'subject' denotes a collective body. It is this contradiction that expansive learning embraces, and I discuss 'subject perspectives' in detail in Chapter 4.

From the perspective of ELT, innovation involves meaningful social action. It is radically different from *knowledge management* approaches, which *reify* innovation as if it is a disembodied thing that stands apart from people. ELT stresses that people have agency and can actively shape their future activities. In other words, people create organizations and not the other way around.

Recent research into organizational change and innovation claims that by using the analytical framework of expansive learning, it is possible to uncover 'the invisible battleground' of interacting agencies and to focus on the aspects of the system that impede or assist collaboration and the reconstruction of 'knowledge' in a new social context (Hasu, 2001; Toiviainen, 2003). 'An activity system is by definition a multi-voiced formation. An expansive cycle is a re-orchestration of those voices, of different viewpoints and approaches of the various participants' (Davydov, 1999: p. 35). An interesting observation about the sustainability of change relates to its perceived value for those involved. As suggested in the previous section, complex change requires people working insightfully on the solutions and committing themselves to concentrated action together (Fullan, 1993: p. 34). This suggests that clarity of communication and shared purpose is vital for success, and arguments and debate are healthy as encapsulated by Davydov. 'The sense of a person's action lies not in the action itself but in its relation to other members of the group' (Davydov, 1999: p. 73). When applying the theory of expansive learning to the organizational learning and to *innovation* in the workplace, Engeström (2000, 2001, 2003) uses the term 'developmental transfer' to describe the collaboration between diverse groups of people at work. At Engeström's Web site at the Centre for Activity Theory and Development Work Research, a major section is devoted to workplace learning and developmental transfer.[10]

The model depicted in Figure 2.4 recognizes the difficulty of developing 'conceptual tools' to understand organizational dialogue, multiple perspectives, and networks (Engeström, 2000, 2001). The concept of an activity network evolved to describe the complicated relationships between multiple activity systems, in which the smallest *unit of analysis* is an activity system. Innovation is described as 'collaborative interaction in which activity systems learn something from each other'. The term 'developmental transfer' distinguishes 'organizational learning in the workplace' from other forms of expansive learning (Tuomi-Grohn, 2003: p. 226).

Engeström appropriated and adapted Vygosky's notion of the 'zone of proximal development' (ZPD), which was originally used to explain how children

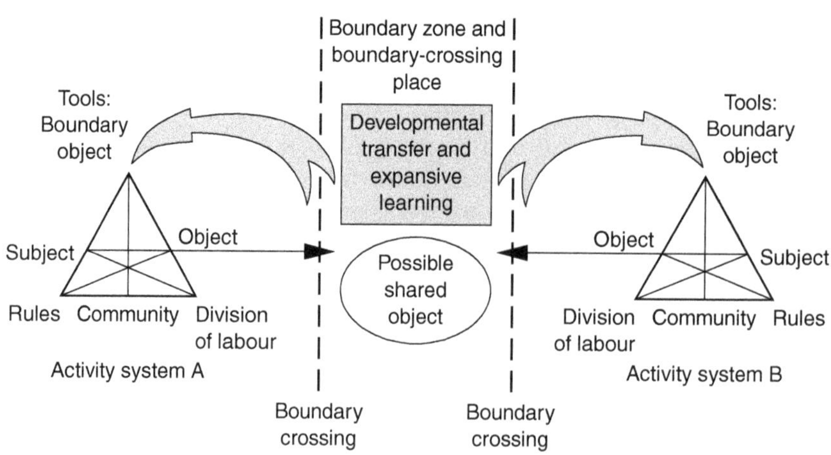

Figure 2.4 Developmental transfer.

acquire a new concept. The ZPD is the ideal space in which a child can learn and represents the optimal distance between the old and new concept. If the gap between the old and new concept is too wide, it is difficult for the child to learn. Engeström adapted this insight to develop the concept of a collective ZPD to explain the optimal distance between routine activity and new forms of societal activity. Organizations can learn new ways of doing things when the gap between old and new practice is not too large. 'The study of an activity system becomes a collective, multi-voiced construction of its past, present and future zones of proximal development' (Davydov, 1999: p. 10). This suggests that ambitious change can engender radical leaps from old to new ways of doing things, and this is very risky. Consequently, studies of knowledge creation in complex organizations focus on the relationships between personal and institutional change.

2.6 The paradox of creative destruction

ELT and complexity theory have common roots in ideas developed by Hegel about the paradoxical coexistence of stability and change. The dialectical logic whereby opposites such as creation and destruction are present at the same time drives the key concepts in both complexity theory and ELT.

> One of the key insights from complexity sciences is that complex interactions of this kind produces at the same time inseparable creative and destructive things ... we are never going to be able to devise a system or procedure which will avoid destructive things happening. The best we can do is try to understand the destructive nature of our interactions and the creative nature too.
>
> (Stacey, 2002: p. 13)

In both cases, the tension between polar opposites stimulates the creation of something new. It is therefore useful to discuss the points where they converge. Stacey refers to the quality and style of interaction. He argues that complexity science adds an important dimension to the idea of human action as self-organization and emergence. 'My experience is that organizations don't become what they become because of their long-term plans.... Other organizations also have long-term intentions and what actually happens will be the intermeshing of all these intentions' (Stacey, 2002: p. 8). Stacey suggests that policy makers have fallen into the trap of thinking that they alone have the power to design the system, and once it has been set up, individual autonomy is constrained by the given design of the system (Stacey, 2002: p. 6). 'For me the greatest insight from complexity sciences is that interaction patterns itself. Coherence and purpose emerge in the interaction between us, in the conflicting constraints that we impose on each other' (Stacey, 2002: p. 11). In complexity theory and ELT, there is a common appreciation of Hegel's view that the individual and the social cannot be separated. This contrasts with the duality of Kantian thinking.

> Indeed, individuals arise in the social, which they are simultaneously constructing. This is clearly a paradoxical or dialectical perspective in which individuals are simultaneously forming and being formed by the social. Hegel's' dialectic method of thought is reflected throughout his discussions of the social evolution of consciousness, self-consciousness and knowledge.
>
> (Stacey, 2003: p. 211)

The idea of an overarching activity network comprising sub-activity systems is conceptually similar to descriptions from complexity theory.

> The notion of system is based on a spatial metaphor of an 'inside' and an 'outside'. A system is a 'whole', and 'inside' separated from an environment an 'outside' by a boundary. This immediately establishes hierarchical levels. At one level there is a system and at another level there is the environment, usually thought of as a supra system, or more encompassing whole.
>
> (Stacey, 2003: p. 279)

From both theoretical perspectives, boundaries are significant. No individual is capable of remaining outside a system of interaction and the power relationships that arise between people in organizations. Although Stacey draws close to addressing the issue of power, he does not develop ideas about how hierarchical relationships between organizations might be understood or mediated.

> It can be argued that all human interaction is power related because we are continually constraining one another and so any change in the way of thinking is going to shift power relations and – presumably this may engender resistance from those currently in the ascendancy.
>
> (Stacey, 2002: p. 12)

Much regional and national policy assumes that organizational partnerships between the public and private sector will be more effective than organizations working separately. However, complexity theory suggests that if the number of partners increases beyond an optimum point called the tipping point, the partnership will fail to function well. At the *tipping point*, an increased number of partners decrease the effectiveness of the partnership because communication becomes too complex and inefficient. The involvement of large numbers of public and private sector partners causes the communications to become unwieldy as the links between partners become too complicated. In practical terms, innovation capacity reaches a maximum at the edge of chaos. However, beyond that point, the capacity deteriorates.

Communication remains failure prone, and this can become acute in complex multi-agency partnerships. Quality control of communication is crucial for developing the knowledge base of the system, but investment in developing good communications between partners is undervalued. Policy makers pay little attention to creating and sustaining long-term relationships between organizations with distinctive and contrasting cultures.

It is clear that partnerships are difficult to manage if they become too large, especially if they include individuals or organizations that do not add value. The inclusion of extraneous partners tends to distort the partnership and reduce efficiency. In terms of complexity theory, a large partnership is likely to reach the tipping point very quickly. If this is expressed in terms of ELT, one could say that there were too many competing objects and inadequate mechanisms for mediating the boundaries between organizational cultures.

On closer analysis, this suggests that to attain its optimum functionality, partnerships should not include too many partners. Each partner must add value, and selectivity is important. The case studies illustrate some contrasts in the development and implementation of innovation policy, which result in partnerships of varying complexity.

Stacey argues that the shift in the relative importance of economic activity from making things to providing services depends on being knowledgeable, being skilful. 'If we put what we think is knowledge down in a book we simply have an artefact. It is only knowledge when someone reads the book and everyone may read it in different way and interpret it differently' (Stacey, 2002: p. 7). Tacit knowledge, know-how, and skilled behaviour are implicit, but they are not easily codified and made accessible to others and ultimately managed by organizations. For instance, Blackler (1995: p. 1026) described important distinctions between *embrained knowledge*, which depends on conceptual skills or cognitive abilities, and *embedded knowledge*, which resides in systematic routines, relationships between technologies, roles, and procedures. This leads to the problem of how to measure this major asset, knowledge, and to understand how to control and manage it. It is extremely difficult to devise accurate metrics for the so-called knowledge economy, and consequently, policy makers succumb to the temptation to record what can be quantified rather than what is truly important. Stacey (2002: p. 7) argues that it is impossible to manage knowledge because knowledge emerges in the interaction between us. Innovation can be understood in terms of how knowledge is created, organized, and diffused in research and development activity. Stacey (1993, 1996, 2003) argues from the perspective of complexity theory that innovation involves unknown outcomes.

From the perspectives of both ELT and complexity theory, policy intervention is risky because the essence of innovation is its unpredictability. Innovation requires complex organizational change and development. When innovation policy is delivered by a multi-agency partnership, emotional and motivational ambiguity is certain. Even in simple activity systems, the object of activity is dynamically evolving. The object is likely to be even more elusive, transient, and contested under the influence of the various organizational partners in a complicated activity network.

2.7 Visualizing the structure and process of activity

Depicting both the structure and dynamics of activity systems presents a continuing challenge for activity theorists. Engeström's most persuasive illustration of the interplay of multiple activity systems describes the development of a new policy to improve the care of chronically ill children. Expansive change occurred when a policy was developed to bring together all those responsible for the child's welfare in a holistic way to improve the treatment by understanding the motivations of the parents and the responses of the medical professions to specific symptoms and events in the child's medical history. If the new practices and the old practices are not too different, they will fall within the collective ZPD and the proposed change is likely to be adopted. The new *patient care agreement* represents an improvement in care for the child and the family and a more coherent approach from the medical care team with less wasteful cross-referrals.

However, it could be argued that the complex interaction of activity systems is not evident to scrutiny because what remains is simply the change in practice. In other words, the policy itself remains, while the description of activity systems affords a complicated way of explaining how the change happened. Figure 2.5 depicts the three activity systems that most affect the healthcare for a chronically ill child and illustrates the complex interactions between three activity systems (Engeström, 2001: p. 155).

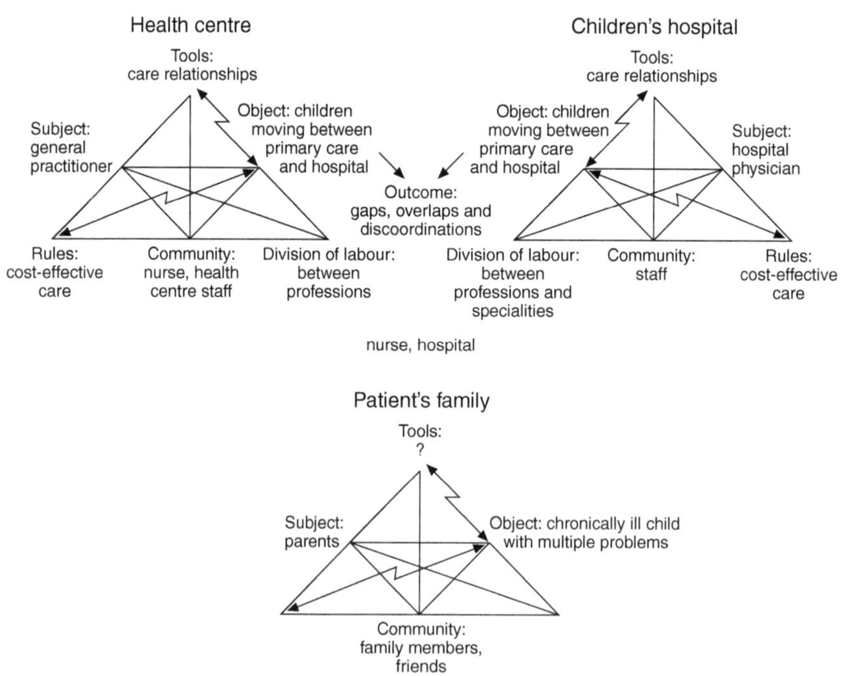

Figure 2.5 Depicting structure and contradictions between activity systems.

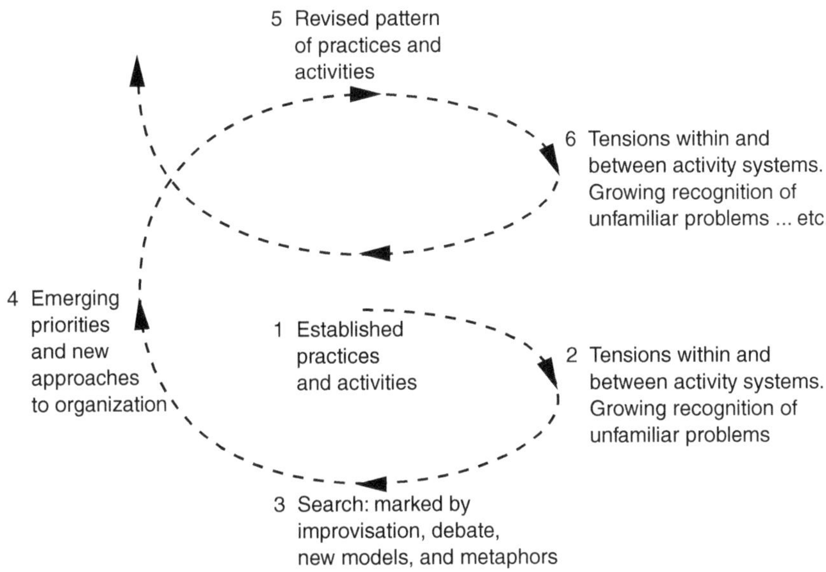

Figure 2.6 Dynamic tensions in activity.

Depictions of interactions between three or more activity systems soon become impenetrable. I found it impossible to use this diagrammatic format to depict the vast array of complex interactions in my case studies. Alternatively, it is possible to depict activity as a dynamic process of change. ELT searches for internal contradictions as the driving force behind innovation and change and analyses the activity historically. Tensions within and between colliding activity systems drive expansion of 'the object of activity'. This is shown in the Figure 2.6 which is adapted from a 'process model' developed by Blackler *et al.* (1999: p. 5).

It is very difficult to depict the structure of activity systems at the same time as conveying a sense of a dynamic process. Activity systems and their interactions are unstable and transitory, and consequently, it is very difficult to capture the continual dynamic flux of activity in graphic form.

The expanded object of activity arising from the collaboration of two or more activity systems represents a temporary cessation of change. From another sociological perspective, the concepts of bridging and provisional stability offer a similar analysis of a respite from change. The concept of provisional stability refers to circumstances and tools that create a temporary degree of stability. It suggests that 'provisional stability' enables people to 'pause for thought' prior to decisions for future action. It offers a temporary solution to changing practices and reduces angst by helping to crystallize a new way of doing something, which is necessary in order to move from potentially destructive instability to a new framework for action (Bonamy *et al.*, 2001; Fullan, 1993). The rapidity of change means that those involved in complex change need to respond creatively in order to survive during periods of change because they are associated with 'chronic' anxieties. Provisional stability describes a conceptual bridge between old and new ways of doing things, and this is similar to the way in which activity theory resolves uncertainties and tensions during the expansion of the 'object of activity'.

Both these theoretical perspectives share a commitment to reflection as an important stage in the process of change. In activity theory, the inherent instability of an activity system is rebalanced when a new expanded object connects the old and new way of doing things. Stabilizing and adjusting the activity system is a constant battle because new dilemmas and tensions keep arising. Like the concept of 'provisional stability', the expanded object can create a 'breathing space' for reflection and a respite from continuous flux. 'People construct their institutions and activities above all by means of material and discursive, object oriented actions' (Davydov, 1999: p. 10). ELT adopted the concept of a boundary object from the sociology of science. Boundary objects are agreed and shared between communities yet satisfy the informational requirements of each of them (Star, 1989: p. 46). They act as tools to link two or more activity systems. 'A boundary object acts as a basis for negotiation and knowledge exchange between differentiated communities of practice or activity systems. Its virtue is its capacity to accommodate local "dialects" ' (Sapsed and Salter, 2004: p. 1519). From this theoretical perspective, a discourse can be created as an outcome of activity, which may also be used as a boundary object. This suggests that discourses can be both global and local and yet be understood to mean the same thing to different communities or activity systems.

2.8 Overview

Expansive learning links a theoretical perspective to an action-orientated framework in which to interpret multi-dimensional relationships between activity systems.

Although it is not 'predictive', ELT draws attention to the 'learning potential' for those engaged in implementing innovation policy. It shaped my data collection and analysis and my overall methodology and research design. This is discussed fully in Chapter 3, which explains how ELT was used as a heuristic device to guide the collection and analysis of empirical data and to interpret the dynamic tensions between old and new practices as they evolved.

Public policy implementation is dominated by the discourse of partnership, and 'knowledge economy' policies assume that partnership can be taken for granted when private and public sector organizations work together. The theoretical concept of *policy borrowing* referred to the implications of adopting a policy that has been successful in a specific regional context and expecting it to achieve the same results when it is transplanted into a different environment.

Chapter 3 describes the methods I adopted in my case studies to show how organizational learning can be visualized. Organizational learning in complex multi-agency partnerships is depicted in diagrams that show cooperation between activity systems in a *collective ZPD*. My next chapter begins with a description of my methodological journey into sociocultural theory.

3 A natural history of my research methodology

3.1 My methodological journey in the rough terrain of innovation

My research started in the midst of experiences and events, which made more sense in retrospect than they did at the time. It is important for me to acknowledge some unevenness in the case studies in terms of their timing, the volume of data, and my access to key informants. As explained in Chapter 1, my work role brought me into the regional policy arena and made me aware of the ambitious policy to develop three geographically defined high-technology corridors (HTCs) in the West Midlands in the UK. The policy concept was borrowed from the USA and mirrored the success stories of Silicon Valley and the clustering of high-technology companies around Massachusetts Institute of Technology (MIT) and Route 128 in Boston. The HTC policy concept, created multi-agency partnerships to catalyse three sub-regional knowledge economies. With hindsight it seems that in the UK case study, I had unique access to people collaborating at various policy levels in the full range of private and public sector organizations. I was able to develop my theoretical insights over a period of more than four years and to extrapolate and validate these insights in my case studies in the USA and Japan.

To return to my starting point in the UK cases, three things intrigued me. First, progress in the HTCs seemed to have stalled, and because many people were involved, it was difficult to work out who to talk to and who took responsibility for the implementation of the policy. Second, at the outset of the Rover Crisis, the Rover Task Force report had been instrumental in the initial development of the policy concept.[1] A great deal of money was spent on consultants who analysed the industrial problems in the West Midlands. Although their reports included recommendations and proposed solutions, there was relatively little time or effort devoted to understanding the implementation process. In other words it was assumed that the implementation of this very complex policy would be unproblematic and so when difficulties arose these were not diagnosed or systematically addressed.

Third, despite the time and money spent on the analysis of regional issues, opportunities to consider feedback from the people implementing the policy were missed. In other words, a reflective evaluation of progress could have created insights and modified the developmental trajectory of HTC policy.

I began my research by talking to people who had influenced the development of HTC policy and who understood its roots in the Rover Crisis of 1999 and the response of the Rover Task Force. I was interested in the anomalies between the theory and practice of using *multi-agency partnerships* as vehicles for developing a regional knowledge economy with sub-regional focal points for technology transfer activities. It was assumed that the policy intention (theory) would lead to the desired outcomes and little thought was given to the nature of the 'implementation process'. I wanted to get an insight into the experiences of those involved in implementing the policy and decided to use sociocultural theory as a theoretical framework for comparison of the development of the three HTCs.

Social practice theory has particular relevance to the operation of technology corridors because the primary 'unit of analysis' is neither the individual nor social institutions but rather the informal 'communities of practice' that people form as they pursue shared enterprises over time (Wenger, 1998). Sociocultural theories offered an alternative to organization theories that focused exclusively on decision-making and on the notion of 'strategic rationality' (Blackmore and Lauder, 2005: pp. 97–98). The rational model ignores the inherent complexity and ambiguity of the social practice dimension of organizations and their environments as discussed more fully in Chapter 2. The interests of stakeholders in organizations strongly influence discursive practices and multi-agency practices do not seem to evolve in any predictable linear way. From a social practice perspective, 'organizational culture' can be defined as a patterned system of perceptions, meanings, and beliefs about the organization which facilitates 'sense-making' and guides the behaviour of people at work (Bloor and Dawson, 1994; Weick, 1995).

I was interested in the way different variants of social practice theory focus on the process of 'sense-making', which is what happens when people make retrospective sense of the situations in which they find themselves. The attention to the social dynamics of organizations suggests that 'sense-making processes' shape organizational structure as well as individual and group behaviour. I was particularly interested in three theories, which were orientated towards studying organizational practices. My initial thinking was influenced by the work of Lave (1993) and Wenger (1998, 2000) and recent applications of communities of practice theory (COPT). This was appealing because of differences in the organizational cultures and practices of the so-called partners in the HTC policy experiment. COPT focuses on small group interactions in the workplace, but the unit of analysis seemed too imprecise. It does not distinguish sufficiently between the centre and the periphery. COPT hints at the significance of power and influence but never really addresses the issue of power even when discussing the interactions between multiple communities of practice (Wenger, 1998, 2000).

I was also attracted to actor-network theory (ANT) because it spoke of technological and human *actants* and their capacity to act in *networks* to drive technological change and innovation (Law, 1992, 2000; Law and Hassard, 1999;

Miller, 1997). ANT recognizes the significance of technology itself as an agent for change and considers practices over a broader canvas and over a longer time frame. Consequently, ANT is concerned with change in whole industries spanning decades or even longer (Fox, 2000). By contrast, COPT is concerned with small group interactions at work and builds upon notions of situated learning; it takes a small unit of analysis in contrast to ANT, which adopts a much larger unit of analysis.

I was, at first, undecided about which one of the relevant sociocultural theories would provide the strongest theoretical basis for my research. However, recent studies of innovation in complex organizations used expansive learning theory (ELT) in very interesting ways, and this convinced me that it was appropriate for my study of multi-agency learning in the HTCs.

My research methodology aims to clarify how 'tools' mediate the interactions of organizations to shape practices related to the enactment of HTC policy. For example, the 'knowledge economy discourse' and the Rover Task Force funding package were 'tools' that became powerful mediators of activity. Discourses can be persuasive; they are tools that embody accumulated social wisdom and can act in two ways to model thinking. Tools have agency and can embody thought and can actively shape thought. Recent research has shown that new tools and technologies can influence our thinking in unpredictable ways (Cole, 1999; Cole and Engestrom, 1993; Cole and Wertsch, 1996; Davydov, 1999; Engeström, 2001, 2004; Engeström *et al.*, 1999; Hasu, 2001; Ludvigsen *et al.*, 2003).

ELT conceptualizes organizational learning as collaboration, which can be seen in the emergence of new practices resulting from working 'across' culturally distinctive activity systems. In studies of phenomena where the boundaries are unclear, the case study method is a well-established approach. It offers flexibility in researching evolving multi-agency partnerships where there is uncertainty about when the activity begins and ends (Yin, 1993: p. 3). Even though the boundaries of the HTC phenomena seemed very unclear, each HTC could be treated as a *meta-activity system* with a *developmental history* that can be described.

Gradually, I appreciated that there were more interesting ways to study the HTCs than as three straightforward *case studies* of a policy implemented simultaneously in different cultural and historical contexts and at different sites. I considered each HTC as a multi-faceted organizational case study.

Although my theoretical insights grew initially from the UK case studies, I have been able to use similar methodologies to analyse two case studies in Japan and two in the USA. ELT does not specify a particular type of data; it suggests that the practical implementation of innovation policy could be understood and analysed from different subject perspectives. In other words, 'innovation projects' can be analysed as discrete experiments where individuals and organizations can interact in new ways. The term 'subject' can refer to an individual person, to a member of a small group, and also to a representative of a large collective community. It is possible to gather interview data from an individual subject, positioned in relation to their membership of particular communities and cultures. Similarly, interview data can reconstruct the location of the subject in terms of the horizontal and vertical interactions between activity systems. (This is addressed more fully in Chapter 4, which reconstructs subject perspectives.)

Subject positioning is depicted in relation to membership of multiple communities of practice and in relation to overarching communities or social worlds. At first, I was troubled by the difficulty of depicting vertical relationships between activity systems, which had varying degrees of power and agency. For example, I began to visualize macro-, meso-, and micro-levels in the HTCs as hierarchically linked social worlds, arrayed like Russian nesting dolls, one inside the other. As my research progressed, I realized that the data showed that interactions within and between levels are even more complicated. (This is discussed at length in Chapters 6–8.)

I thought it seemed fruitful to analyse the cases in various ways. I decided to make a multi-faceted study by tracing the experiences and changing practices described by individual subjects in interviews. It was possible to locate individual subject perspectives on a two-dimensional matrix with a vertical axis and a horizontal axis.[2] Interviews could capture the subject's perspectives in relation to the level at which they were located in the policy implementation structure. This data could also co-locate their position in relation to a particular organizational culture.

It is important to appreciate that the complexity of the UK case studies forced me to consider the tensions inherent in the positioning of an individual respondent or subject. I realized that I could focus on three aspects of interaction and collaboration. First, I could look at variation across different sites based on the geographical focus of the HTCs. Second, I could consider the cases as studies in terms of variation in stakeholder practices. Third, and most importantly, I could analyse the data to reveal differences in the practices adopted at the three levels of policy enactment. I believe this awareness of the complexity of subject positioning influenced the approach I adopted in my case studies in the USA and Japan.

Although the case studies were complex and multi-faceted, it was possible to treat each innovation project as a *meta-activity system*, in which several sub-activity systems interacted horizontally across organizational cultures. ELT describes this as *polycontextuality* such as collaboration between experts to tackle a multi-disciplinary problem.

ELT offers a stronger analytical framework than COPT because it focuses on a dynamic transition in learning by individuals embedded in activity systems in pursuit of an object. Nevertheless, it does not resolve questions about the impact of power differences between activity systems. It cannot depict as 'collaboration' the conflict between competing objects and discursive practices in the vertically arrayed levels of policy enactment. The dynamics of these *vertical interactions* are characterized by differences in agency and power. Interactions between relatively powerful and weak activity systems cannot be described in the same way as *polycontextuality* because they are different in quality.

I experienced a good deal of anxiety about using an evolving theory. During the course of the research, I began to realize that although ELT had been used in studies of complex organizations, it did not offer a robust explanation of issues of power and agency. This was a gap in ELT, which I decided to pursue in my research, by pushing the theory to its outer limits. I have outlined these complexities and limitations at the outset, so that the claims for the data analysis are seen within this context.

3.2 My research design

My approach to data collection is consistent with the conception of knowledge as a social construction, which is dynamically created and reconstructed as individuals and groups interact. The research questions, stated in Chapter 1, focus on emergent collaborative practices and link the heuristic device of ELT with the data collection and analysis. In this way, my research design links the theoretical methodology and ontology with the data analysis, so experiences described by interviewees were captured and represented in transcripts as summarized in Table 3.1.

I used ELT to depict the developmental history of the innovation projects. The interconnectedness of all the elements in an activity system suggests that by understanding an individual subject perspective, it is possible to construe their relationship to other elements in the system. In other words, the elements within an activity system are in dynamic tension, and the process of innovation brings several different activity systems into new relationships with one another.

Table 3.1 The context and scale of the case studies

Case studies	*National policy context*	*Regional innovation programme*	*Local projects*
USA ⇓	Market-led economic growth and competitiveness	Silicon Valley	Role of venture capitalists Context for financial and service industries
	Market-led economic growth and competitiveness	California energy	Zero-energy homes Public interest research
UK ⇓	DTI competitiveness and innovation policies and EU Lisbon Agenda	HTC1 innovation programme HTC2 innovation programme HTC3 innovation programme	Numerous innovation projects of varying complexity. Focus on manufacturing engineering, ICT, and medical technologies
Japan ⇑⇓	National innovation policy, clustering, and emergent high-technology sectors	Mature textile industry Photonics	Design-led innovation projects Local R&D clustering Projects

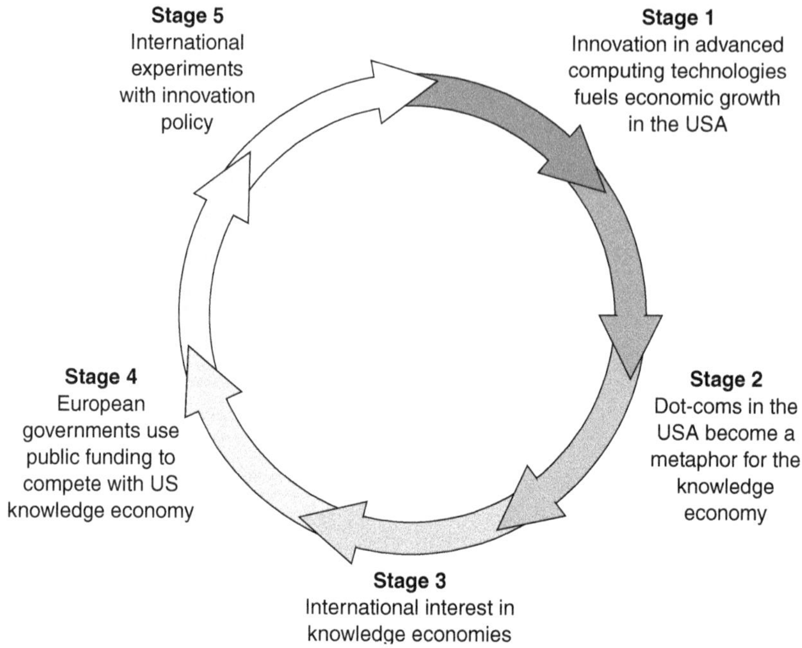

Figure 3.1 A knowledge-economy policy-borrowing cycle.

Policy experiments and their practical impact can be shared more effectively to maximize the potential to learn from what worked well in particular circumstances. Figure 3.1 depicts the relationships between the case studies in terms of a policy-borrowing cycle in which an experiment in one national context influences experiments elsewhere.

3.2.1 The UK case studies: HTCs

The UK case studies show how the implementation of HTC policy was a direct result of the crisis at the Rover Automotive Plant in 1999, when it was threatened by possible closure with the potential loss of 27,000 jobs. The crisis was a catalyst for the region to change from dependence on traditional manufacturing by developing a regional 'knowledge economy'. The policy was influenced by the intention to map research expertise in the universities and use this to attract new companies or persuade existing companies to relocate in HTC areas.[3] The Rover Task Force devised an ambitious policy for technology transfer to be implemented in geographically defined sub-regions called HTCs.

3.2.2 The Japanese cases: textiles and photonics

By contrast, the Japanese Government recognized a need for a national innovation policy after the Second World War in 1945. Economic recovery was driven by an enduing alignment of private sector companies with the Government to achieve economic growth through innovation.

Japanese textiles have a rich historical tradition of silk and cotton kimonos. Consumers valued skilled manufacture of fabrics and individual design and were willing to pay a high price. However, a national shortage of natural resources such as cotton and wool influenced the trajectory of innovation in the textile industry. During the past twenty years, it stimulated the inventive use of synthetic fibres in what was perceived to be a mature and declining sector. My first Japanese case study explores design-led innovation in the textile industry by focusing on the collaborations initiated by a small Tokyo-based textile design corporation. Textile manufacturers and designers began to work ever more closely to respond to increased competition from overseas and changing consumer tastes with innovative products.

Recently, economic policies have moved their focus from mature and declining industries towards new high-technology sectors with rapid potential to grow. My second Japanese case study examines the process of innovation in new industry sectors producing high-technology products and services. Government policy has been designed to encourage public–private sector partnerships and to support multi-organizational networks. The Japanese Government encourages collaboration on the costs of research and development by making grants available to networks of companies using new technologies such as photonics.

> In order to avoid the useless duplication of research work and to push forwards the development and application of research efficiently to needs in industry, it is preferable to establish a study system for joint research (by) industrial and academic circles. Japanese research money is spent more efficiently because of this strong cooperative research effort.
>
> (Eagar, 1986)

In recent years, Japanese innovations in photonics have been extremely successful. The Japanese Government has devised coherent policies to support research and development in industry and academia.

> a higher proportion of business collaboration with universities and research institutes results in novel products than collaboration of other kinds. Partnerships with universities and research institutes are thus a useful form of collaboration for SMEs developing new products, and such partnerships are expected to play an increasingly important role in the future.
>
> (JSBRI, 2005: p. 67)

In particular, new forms of business collaboration models are thought to be effective and exemplify the ‘triple helix’ approach to innovation policy. This fuses technologies and know-how across different research disciplines and industry, which helps to escape the conventional mindset of individual enterprises and researchers.

3.2.3 The US cases: Silicon Valley and sustainable energy innovation

Two case studies illustrate a diversity of collaborative processes that have been successful under specific financial and cultural conditions. First, Silicon Valley has inspired other regions and nations to try to replicate its success. Outside the USA, other national governments used public funding to catalyse the clustering of high-value companies creating new technology products or services. There are numerous examples of such experiments in 'policy borrowing'.

The 1990s were marked by the rapid growth and instances of spectacular failure of a group of new Internet-based companies commonly referred to as 'dot-coms'. A combination of rapidly increasing stock prices, individual speculation in stocks, and widely available venture capital created an exuberant environment in which many of these businesses dismissed standard business models, focusing on increasing market share at the expense of the bottom line. Google is an example of a dot-com company that was created from a university spin out which successfully weathered the financial storms of the dot-com crash and transformed itself into one of the most successful US corporations today.

My second US example is the private and public sector interest in sustainable and renewable energy by new policies in response to the energy crisis in California.

In the USA, the energy crises in California 2001 resulted in a reassessment of the dependence on conventional fuel sources. Enron cynically manipulated the energy market to create an artificially high price for energy, from which they drew huge profits. The experience caused individual householders to think more deeply about renewable energy that could make them less dependent on state and national corporations.

> Renewable energy is incredibly popular – a Public Policy Institute of California poll earlier this year showed that 83 percent of adults interviewed supported more government spending to boost renewable energy. The state has plenty of sun and wind – experts suggest the Tehachapi region could generate enough wind power to light 3 million homes.
>
> (Martin, 2006)

Although action may have been sparked by the manipulation of energy markets and the eventual collapse of Enron in 2001, there is real commitment to reduce carbon emissions and move the state towards environmentally sustainable energy production and use. The case study of the Sacramento Municipal Utility District (SMUD) illustrates the synergistic activities of the private and public sector in California.

I gathered my primary data in interviews in which respondents described their 'stories'. I used respondent triangulation to identify consensus and contradiction. I analysed the interview data to uncover the 'interests' of individuals and their 'organizational positioning' (Ball, 1994). The interview transcripts positioned the respondents within the policy process, both explicitly and implicitly (Ball, 1994; Troyna, 1994). I used a snowball sampling technique, and I began with a sample of strategic actors who had a major influence on innovation projects. Initial respondents in the first wave of interviews led me to other interview subjects.[4] On one level, I treated the interview transcripts as strong realist stories, and on a second level, I considered them as *discourses*, as sophisticated 'representations' rather than 'guileless descriptions' of events.

Secondary data comprise public documents such as policy statements, reports, and Web sites. Data included international, European, national, regional, or state policy documents on innovation. In Japan, national policy documents and white papers were rich sources of data (Byrne, 2006; IARPS, 2006; JETRO, 2004; JSBRI, 2005; NEDO, 2006; NSF, 2006; United Nations, 2006). In the USA, national and regional energy policy documents and local Web sites provided excellent summaries of the public interest debate (San Francisco Chronicle, 2006a, 2006b). For instance, in the UK, the Department of Trade and Industry (DTI, 2002, 2003a, 2003b, 2003c) and Department for Education and Skills (DfES, 2003a, 2003b, 2003c) publish policy documents on innovation and competitiveness (EU, 2003). It also included the Regional Innovation Strategy, the Regional Economic Strategy, and the Agenda for Action (AWM, 2003; AWM and AfA, 2002; AWM and WMES, 2003; UUK, 2003; UUK and HEFCE, 2001; WMiE, 2003; WMiE and ESF, 2003; WMRA, 2003).[5]

This secondary data proved to be a valuable reflection of different 'organizational' perspectives.[6] Texts such as strategic plans or Web sites were analysed to trace the origins, nature, and structure of the discursive themes that influenced the policy. In this sense, the evolving discourses were interpreted as both outcomes and tools of activity (Prior, 1997). In some instances, the new discursive positions were compared with previous and separate positions. Table 3.2 summarizes the research design and data collection and analysis cycles. It explains how the theoretical drivers shaped my research methodology.

Relationships between the case studies were considered in terms of 'policy borrowing'. Successful regions such as Silicon Valley acted as exemplars that Governments attempted to replicate. However the environmental and economic contexts of these experiments varied enormously. It was by no means certain that it was possible to stimulate regional knowledge economies by injecting public funding. Figure 3.2 illustrates some of the connections between the case studies and the opportunities for practice to inform theory. It shows the cross-pollination of policy ideas and the global influence of the Silicon Valley phenomenon.

Throughout the data collection and analysis cycles, I considered the data in various ways and looked for patterns and anomalies in:

- tensions between elements of activity,
- tensions across activity systems at the same levels (horizontal), and
- tensions between activity systems at different levels (vertical).

Table 3.2 Research design including data collection and analysis plan

	Timing	*Data collection*	*Data analysis*	*Data exposition*
Phase 1	Winter 2003	Semi-structured interviews with eight key individuals	Identifying patterns and themes in the data. Mapping tensions between rules and divisions of labour as they emerged from the data	Initial matching of data against elements within an activity system. Initial analysis of interviews as stories, representations of interests, and discourses using the implementation staircase model to think about policy implementation
Phase 2	Spring 2004	Semi-structured telephone interviews with HTC steering group members including public and private sector stakeholders (fourteen interviews)	Searching for relationships and patterns in the data. Using Atlas ti software to map the tensions between elements of activity and between activity systems	Initial analysis and reconstruction of practices in the HTCs as complex activity networks of embedded activity systems
Phase 3	Spring and summer 2004	Interviews to fill gaps in data or explore ideas about tensions between ‘espoused and tacit objects’ (five individuals)	Matrix analysis of major themes in data within and across HTCs. Analysis of practice and development of propositions using ELT as the heuristic tool Identifying competing objects at different levels of policy enactment	Initial typologies of variations in data from respondents positioned differently in the enactment of policy Exposition of links and boundaries between activity systems. There were three levels, the macro (top regional policy level), the meso (sub-regional levels), and micro (the local projects level)

Phase 4	Summer, autumn, and winter 2004	Interviews with (five) SMEs or HTC project staff to illuminate 'the facilitation' of dialogue between developers and users most effective in changing practices within SMEs	Analysing the data to illuminate the role of technology translation projects and the use of boundary objects and mediators	Cross-checking tentative findings and constructing an explanatory framework in terms of boundary crossing and expansive cycles. Writing up the conclusions
Phase 5	Summer and winter 2005	Collection of secondary data for US and Japanese case studies	Analysis of international, national, regional, and State policy documents	Testing data against UK findings Checking patterns for inconsistencies and anomalies
Phase 6	Spring–autumn 2006	Collection of primary data via questionnaires and interviews for US (Silicon Valley and sustainable energy) cases. Also Japanese (textiles and photonics) case studies.	Analysis of policy documents. Focusing on SMM innovation practices	Comparing international policy intentions and outcomes. Cross-referencing evidence of policy borrowing
Phase 7	Winter 2006	Checking transcripts with interviewees for US and Japanese cases	Cross-checking final analysis of data	Checking findings and writing up conclusions and theoretical insights

Source: *Miles & Huberman* (1994).

Notes
ELT, expansive learning theory (a heuristic tool and an object of study); HTC, high-technology corridor; SME, small- and medium-sized enterprise (employing less than 250 people); SMM, small- and medium-sized manufacturer (Japanese textiles case study).

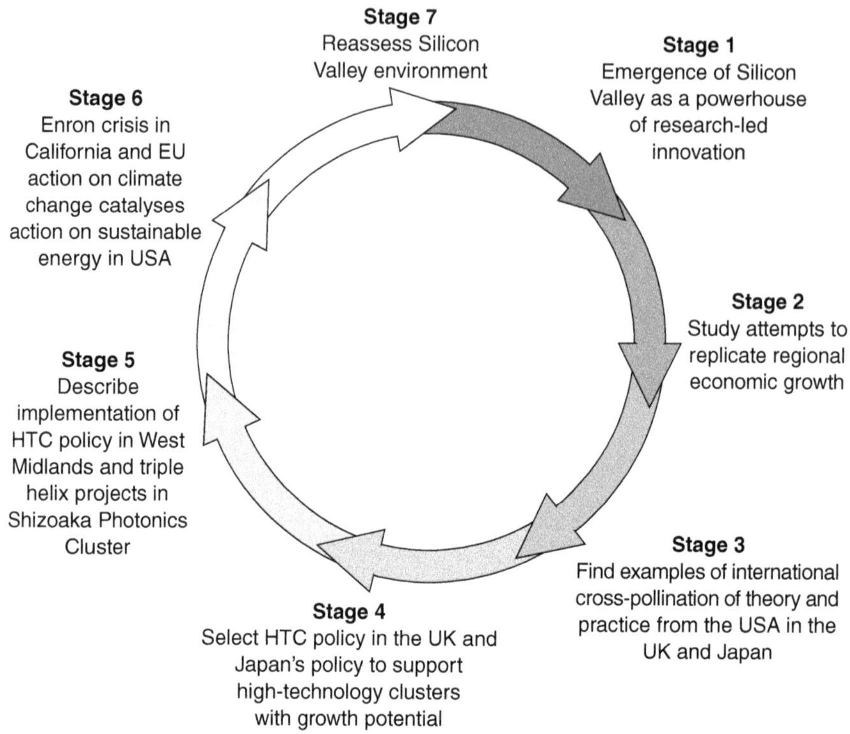

Figure 3.2 Tracing the policy-borrowing cycle in case studies.

To analyse these tensions, I extended an analytical device used by Blackler *et al.* (2000: p. 290) to summarize the patterns in the data and emerging forms of expansive learning in the HTCs. A generalized example of the analysis is summarized in the Table 3.3.

3.3 My method of data presentation

I have not presented the data as separate case studies; instead, it is integrated into thematic discussions and analysis within the narrative. I present my data in vignettes to evoke a vivid impression of actions and interactions. Vignettes integrate quotes within the text to ground the theoretical analysis and to illustrate the interpretation of the empirical data. The vignettes reconstruct the experiences described in the interview data, and they depict key situations and interactions 'in context'.

It is important to clarify that I did not use vignettes as a data elicitation technique. I did not need to use vignettes to collect data as described in studies where researchers deal with vulnerable interviewees and highly sensitive subject matter such as in cases of abuse and violence.[7] I have used vignettes simply as a presentational device to illustrate the complex theoretical implications of

Table 3.3 Emerging forms of expansive learning in the HTCs

Analysis of HTCs	*Activity networks at three levels*	*Multi-voicedness*	*Developmental history*	*Contradictions*	*Expansive cycles*
Who are learning?	Public and private sector partners	There are tensions between organizational cultures and voices Collisions between activity systems of 'partners'	Individuals and organizations are learning to work together at all three levels	Tensions between the rules governing each meta-activity system correspond to division of labour at three key levels	Learning across levels or polycontextuality happens at various times in different levels
Why do they learn?	Tensions between activity systems drive change and collaboration	Dominant discourses prevail	Discourses change over time and the dominance of these varies	An object is created at each level and these three objects compete with one another	Tensions between objects at three levels drive collaboration between levels
What do they learn?	They value the complementary perspectives of individuals and each partner organization	They find shared solutions to problems	They learn to take more risks and to trust the other partners	Tensions between old practices and new practices	They appreciate that they can do more together than by working independently
How do they learn?	Individuals move back to influence their home organization	Dialogue and argumentation drive learning	Boundary objects help partners move across activity systems	Mediators can help learning between levels separated by agency	Confidence in the new practices grows and expands into a cooperative partnership

Notes
Each innovation project was conceptualized as an activity network or meta-activity system comprising sub-activity systems. In the most complex examples, knowledge transfer policy was enacted at three levels, macro, meso, and micro. These levels corresponded to national, regional, and local contexts in which an activity system is the smallest unit of analysis.

unfolding activities as they occurred in a particular location at a particular moment in time.

Even the most recent applications of ELT in studies of innovation focus on polycontextuality or collaboration between activity systems operating at the same level. This gap in theory and the absence of a precedent for my research make an integrated analysis in two directions unmanageable. Consequently, a horizontal analysis of interaction precedes a vertical analysis, and these two analyses made in parallel are then combined at a later stage.

Inevitably, discrepancies arose and contradictory data are included in thematic discussions throughout this book. Nevertheless, new insights emerged which revealed tacit assumptions in innovation policy and the inherent contradictions between theory and practice. My aim was not to propose an ideal recipe for innovation but to describe the impact of the practices that people adopt in different circumstances.

3.4 Discussion of ethical and political issues

Although it was not my intention to conduct an evaluation, I adhered to ethical and procedural guidelines published by the UK Evaluation Society. In the interview data, many of the respondents gave frank and 'politically sensitive' perspectives and interpretations of events and practices. I made clear the academic purpose of my enquiry and assured the respondents that the data would be treated in confidence.

The interview data include transcripts containing 'opinions'. For instance, in the UK case study, some individuals consistently adopted the 'organizational perspective' of the Regional Development Agency (RDA) and used the rhetoric of their employer. Sometimes, opinions were given 'off the record', and these revealed some interesting tensions within the organization. In the US cases, interviews captured the various 'perspectives' of private sector partners and stakeholders who were willing to express opinions on innovation policy. In one of the Japanese case studies, a highly innovative textile design company describes their activities in terms of *artistic creativity*. Collaboration between designers and small- and medium-sized manufacturers (SMMs) is described in terms of synergy and mutual benefit. However, key people in the textile design company always talk of 'learning as an individual', 'learning as a company', and 'learning by working with SMM partners'. So, this vision of what they do is consistent with the way in which ELT defines innovation as collaboration.

Although it is important to protect the confidentiality of the data, there is a corresponding need to attribute remarks to the speaker and to place the individual perspectives within a context of organizational positioning. For instance, the empirical data captured clear divisions of labour within the RDA, in which two different sets of rules and cultures operated. The visionary and inspirational role in the development of policy can be seen to be in tension with their practice of rigorous monitoring and fiscal management of HTC projects, and this is explored more fully in Chapter 8.

In order to set realistic claims for the 'credibility' of the primary data, I shared the interview transcript with each respondent. They were asked to check the text for accuracy of content and interpretation. Notes of conversations were likewise checked to resolve ambiguities, and in some instances, I interviewed a respondent more than once. Sometimes, I framed the issues or the question slightly differently to detect changes in their view perhaps influenced by reflection on events. My methodology recognized the possibility that a respondent over a period of months may have reconsidered a 'taken for granted' position, as a result of an earlier conversation. Tracking changes in 'the views expressed' over time was consistent with the social constructivist orientation of the enquiry. This strategy aimed to strengthen the transferability and confirmability of the data (Bauer and Gaskell, 2000; Kress and Van Leeuwen, 1996; Mason, 1996; Silverman, 2000). I also used respondent triangulation within and across the cases and between primary and secondary sources of data, to ensure the 'credibility' of the data.

Much research celebrates the distance between the researchers and respondents. Some researchers associate this issue with notions of 'objective truth' and get involved in problematic issues such as the relative legitimacy of their truth claims (Beach, 2003). However, my research uses a range of tools to check and validate data. A good deal of effort has gone into checking the 'constructions of meaning' and assuring the provenance and authenticity of the primary data with the interviewees. All the respondents had an opportunity to check the accuracy of their interview transcripts. They were invited to inform me of any misinterpretations of their comments and to make additional observations. My intention was to strengthen 'confirmability' of the data in this study (Lincoln and Cannella, 2004).

After each data collection cycle, my analysis identified patterns and gaps, so that the next cycle of data collection was informed by puzzles or interesting themes as they emerged. I triangulated the interview data across levels and between levels as a method of internal validation. I also triangulated the primary interview data with secondary data.

My approach contrasts with the strategies adopted by Finnish originators of ELT. In some studies of organizational innovation conducted at the Change Laboratory in Helsinki, the environment was structured to enable the researchers to reflect findings back to the participants. Under these conditions, researchers intervened to shape the desired change and claimed that 'communicative validity involves testing the validity of knowledge claims in a dialogue' (Hasu, 2001). However, dialogue with respondents was not realistic or practical given the naturalistic setting in which this research was carried out. There were sensitive issues concerning the expenditure of large sums of public money. Also, it was too difficult to share my interpretation of the data with respondents because the terminology and abstractions in ELT are central to the data analysis and yet very difficult to communicate (Collins *et al.*, 2004). This issue is discussed in Chapters 9 and 11.

To communicate the complex findings of the enquiry, stories with rich contextual detail but limited attribution were constructed from the empirical data.

The research sought to strike a balance between protecting the anonymity of respondents and presenting the data in its 'context'. There is some risk that a reader with local knowledge may have sufficient background information to identify respondents despite the disguise of the referencing system. It has been difficult to resolve this dilemma because comments selected from the interview data only 'make sense' when considered in the context of 'their organizational positioning'. In some instances, comments and 'interpretations of events' are critical of various agencies and are therefore 'politically sensitive'. This may be a potential weakness and limits the capacity of the research to be credible and informative to the key respondents and other practitioners.

An ethical dilemma became evident in balancing the need to contextualize and validate the empirical data and, at the same time, to protect the anonymity of the respondents. The need to resolve this problem of attribution became most acute when referencing the data in the following chapters. Consequently, I adopted a referencing system to ensure that sensitive and confidential comments were anonymized while retaining sufficient empirical data to embed it within its context. In the UK case studies,

- category A refers to an interview with a strategic actor in regional innovation policy,
- category B refers to an interview with an actor at the meso-level, and
- category C refers to an interview with an actor at a micro-level responsible for practical implementation of innovation policy.

In the US case studies, my respondents agreed to be named although I offered to anonymize responses as shown below:

- category D refers to an interview with a strategic actors in the development of state policy,
- category E refers to an interview with a private sector stakeholder, and
- category F refers to an interview with a public sector or community stakeholder.

In the Japanese case studies, it was important to provide contextual detail, and fortunately, my respondents also agreed to be named although I offered to anonymize their responses as shown below:

- category G refers to an interview with a strategic actor in the regional innovation policy,
- category H refers to an interview with a CEO of a private sector corporation (or published interview data from a source approved by the CEO), and
- category I refers to an interview with an associate in a private sector company.

Where the respondent is anonymized, the quotations have a footnote bearing an identification number that links it back to the individual respondent if they preferred not to be named. Each number is prefaced by the code 'PD', which refers to a primary document containing the interview transcript located in a qualitative data analysis software programme.[8]

3.5 Overview

The research design recognizes that ELT proposes an explicit focus on the interaction between actors and their surrounding environment. Activity is a sociocultural construction, and the meaning of activity in any particular context is shaped in relation to the activity systems in which it is embedded.

My next chapters draw on the empirical data to characterize patterns in the relationships and practices. Chapter 4 reconstructs the subject perspectives of key individuals and agencies. Chapter 5 describes interactions across the cultures of the various agencies and focuses on the horizontal relationships between users and developers in the process of innovation. Chapter 6 depicts hierarchical interactions between relatively powerful people at decision-making levels and the practitioners who implement policy but who are at a relatively weak position. Vignettes describe mediation by boundary objects and mediators. My data analysis illustrates that tensions arise from differences between the partners in terms of power and agency. In other words the participating organizations or partners are not equally influential. Systematic data analysis reveals competing objects of activity and the tensions between them.

The vignettes in Chapter 7 focus on collaboration and expansive learning in very complex innovation projects and programmes. Mediation is discussed with reference to different contexts for organizational learning in which people acted as brokers between relatively powerful and weak activity systems.

Chapter 8 focuses on the possible uses and misuses of evaluation. It explores how continuous evaluation supports an ongoing learning process, which can inform adjustments to the trajectory of policy. Chapter 9 summarizes the insights from the research findings.

Chapter 10 proposes a toolkit that can be used to run a health check on a partnership or multi-agency project. It suggests how these tools can be used to detect potential problems and suggest areas where special effort would be fruitful. Finally, Chapter 11 concludes with a discussion of the risks and limitation of the book and a suggestion for future developments of ELT.

Turning now to Chapter 4, it draws on the empirical data to explore the issues of subject perspective and organizational positioning and its impact on strategic decision-makers and on the practitioners responsible for implementing the policy.

4 Reconstructing subject perspectives

4.1 The complexity of subject positioning and identity

A subject can speak as a member of a community associated with a specific work culture and 'perspective' or as a representative of an organization positioned at a particular level of policy enactment. Identities are often in flux and influenced by the experience of working with other people who challenge habitual ways of seeing the world. In sociocultural theory, the 'object of activity' encapsulates the motivation and goal to which people are orientated. Various objects of activity emerge and compete because contradictory sets of rules operate for different people in particular contexts. Also the object changes over time, and so the goal of the innovation process is constantly evolving. The object of an organizational partnership is especially elusive because the perspective of each partner contributes competing 'objects of activity'. The vast array of partners of different types makes it more difficult to expand the object of activity through the synthesis of all these complex and elusive goals.

I found the task of identifying *subject perspectives* very challenging because interview *subjects* are positioned in two dimensions. A subject is positioned along a horizontal axis where different organizational cultures interacted. The same subject is also positioned on the vertical axis at macro-, meso-, and micro-levels of high-technology corridor (HTC) policy implementation. In complex innovation projects, I found it helpful to visualize the location of the perspective of the 'subject' in these stories as points on a graph with the x-axis representing the horizontal collaboration between partner agencies and the y-axis representing the hierarchical positioning of the subject in terms of the agency.

The identity of an actor in one activity system is influenced by their identities in other contexts. The accounts that respondents gave of events and practices were treated as discourses that embodied organizational values, cultures, and attitudes corresponding to particular locations on the vertical and horizontal axis of activity, which influenced their overall identity and *subject perspective*. The methodology aimed to ensure the interpretation of interview data was credible and transferable to other contexts.

The object of an organizational partnership is especially elusive because it emerges from the interaction of competing 'objects of activity' arising from the various perspectives of individuals, groups, and organizations. The vast array of partners of different types makes it more difficult to expand the object of activity through the synthesis of all these complex and elusive goals.

Multiple memberships of different activity systems made analysis of interactions between related activity systems a complex task for which no precedents were available in terms of how to deal with multi-faceted and multi-level analysis and interpretation.

The elusive and changeable nature of the 'object of activity' is influenced by the location of the 'subject' on a horizontal axis, characterized by the cultural perspective associated with each of the 'partner organizations'. At the same time the 'subject' is also positioned on a vertical axis (corresponding to the three key layers of decision-making in the implementation of the HTCs). Consequently the 'object' is influenced by both the vertical and horizontal positioning of the 'subject'. In this sense the 'subject' has a unique perception of the 'object of activity' depending upon who they are, where they are positioned in the implementation process, and when they were studied.

All this changes over time throughout the implementation process. In other words, horizontal and vertical dynamics influence *who* (the self-identity of the subject), *where* (the context of the subject's engagement and their organizational positioning), and *when* (the position of the subject in the chronology of the development of the HTC). The study traces the evolution of individual and collective conceptions of the 'object of activity'. Throughout the following chapters, interactions between complex institutions as they enter into relationships with other organizations in the enactment of HTC policy are visualized as collisions between activity systems and activity networks. These create new dilemmas that continually reshape the object of activity. 'Vertical complexity relates to changes at individual, local and regional community levels, and the interactions between these levels. A host of other political, demographic and geographic factors also applied' (Connell *et al.*, 1995: p. 2). The vertical interactions are expressed as being at macro-, meso-, and micro-level, to distinguish between three different levels of decision-making about HTC policy.

The enactment of a complex innovation policy is characterized by a national, regional, and local dynamics. These interactions are visualized as layers in a sandwich or cake in which the top layer has more agency than the bottom layer. The horizontal relationships are expressed in terms of interactions between different organizations where there are no distinctive hierarchies in agency but significant variations in culture and expertise. 'Horizontal complexity is about working across systems or sectors and agencies and it is difficult to measure this complex array of activities, which have social, economic and physical dimensions' (Connell *et al.*, 1995: p. 2). The 'object of activity' relates to what people understand to be 'the project in hand' (Blackler and Kennedy, 2004). Therefore, the object is described differently depending on the position of the individual in three dimensions. Positioning relates to organizational culture, the level of policy enactment, and the point in time. It was very difficult to be precise about the 'object of activity' because of the inherent difficulty of grasping the meaning of an object to the various 'subjects' at any single instant.

Although expansive learning theory does not suggest or prescribe any particular type of data, it does offer a heuristic framework to set individual actions within a broad social context of stakeholder interactions in the innovation process and knowledge transfer environment. It is possible to think of policy as a story outline adapted at the operational levels to make it 'work in context'.

In a sense, an innovation project is a discrete experiment and can be visualized as a *meta-activity system* composed of sub-activity systems. A project is more appropriately described as a network rather than a system because of the scale and complexity of the interactions. By treating each project as a complex network of activity systems, it is possible to use expansive learning theory to analyse the relationships within and between activity systems 'through the lens' of specific stakeholder perspectives. The research methodology recognizes the risk of imprecision in the units of analysis and the tendency in recent studies of innovation in organizations to use the word *subject* to refer sometimes to an individual, a small group, or a collective body such as an organization (Blackler, 2000; Blackler and Kennedy, 2004; Engeström, 1999; Tuomi-Grohn, 2003).

The vignettes depict subject positioning in relation to membership of multiple communities of practice and in relation to overarching communities or social worlds. The Japanese case studies involved small partnerships and discrete projects. Consequently, it was reasonably easy to depict subject perspectives in the context of the whole activity system as shown in Figure 4.1.

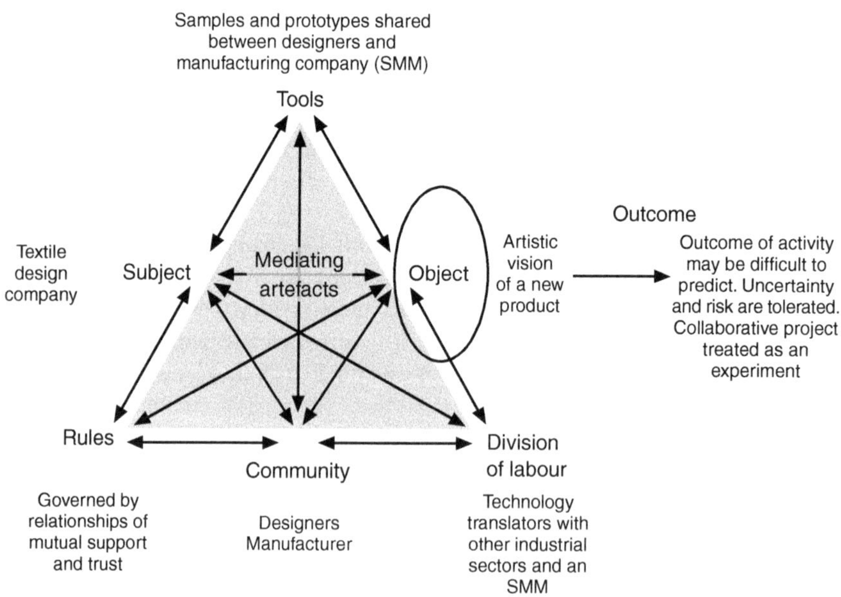

Figure 4.1 Small-scale textile innovation project.

The US case study of sustainable energy involved a more complex network of relationships. For instance, this community project involved a broad range of private and public sector and community organizations in long-term policy. The Sacramento Municipal Utilities District (SMUD) was able to put forward a key actor to mediate the relationships of these diverse organizations throughout a series of practical endeavours as shown in Figure 4.2.

By contrast, the UK case involved a very complex public sector-led partnership involving numerous stakeholder organizations. Although projects could be analysed in a similar way, the projects were also situated within an overarching policy implementation structure with vertical dynamics. Consequently, a subject could be positioned in two dimensions. In the horizontal dimension, subject positioning was determined in terms of their expertise and role. In the vertical dimension, the subject was positioned in the implementation hierarchy, which carried with it a certain degree of agency and power.

There were three levels of policy enactment, each one governed by different sets of rules. Commitment also varied because at each level a representative of an organization is also a unique individual with different personal enthusiasms and preoccupations as shown in Figure 4.3.

At the macro-level, 'HTC policy' was somewhat abstract and described the development of a regional knowledge economy. A strategic vision of innovation and diversification is the espoused 'object of activity'. This was reiterated in published policy documents and reports, and it was as close as policy makers

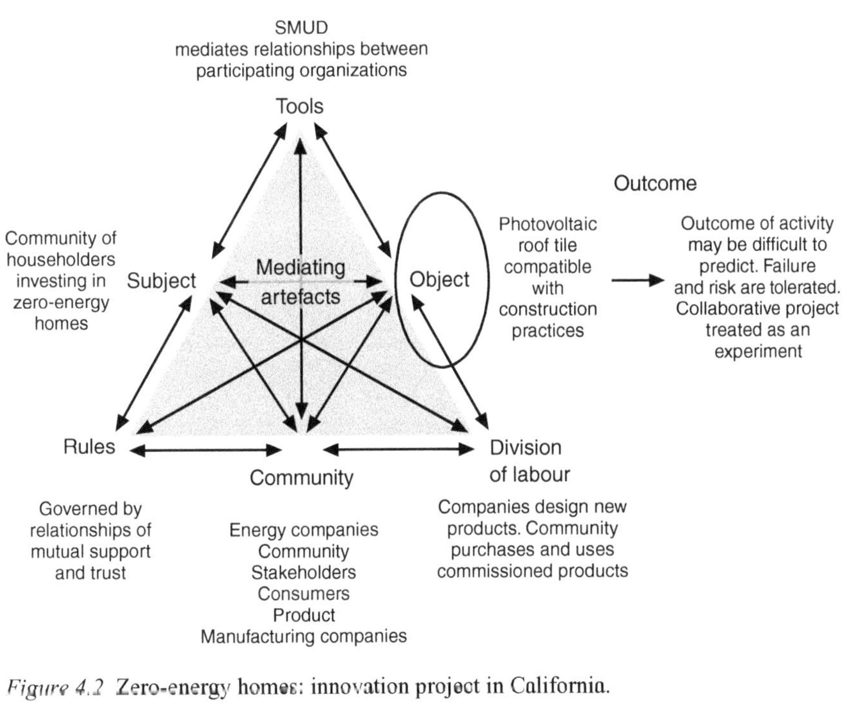

Figure 4.2 Zero-energy homes: innovation project in California.

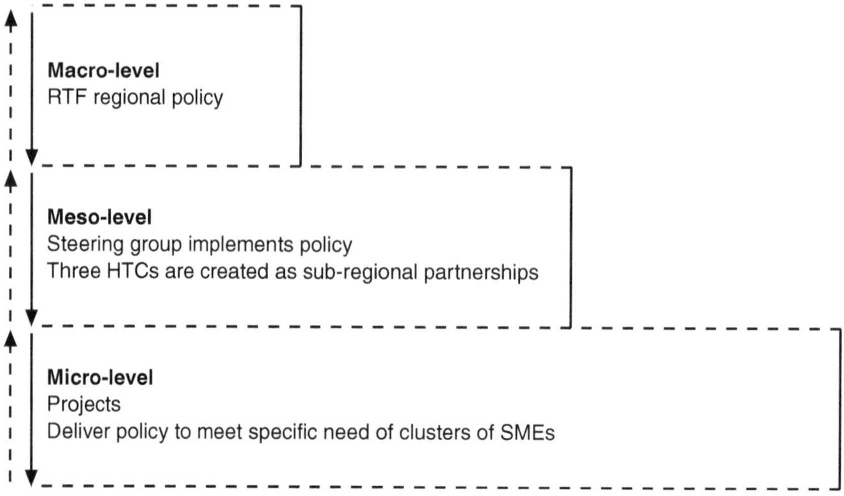

Figure 4.3 Hierarchical implementation of HTC policy.

could get to theorizing about their intended outcome. The focus of effort was orientated towards regional development of a knowledge economy.

Figure 4.4 illustrates my process of data mapping and shows how the model of a *meta-activity system* is used as a tool for thinking about the macro-level data (Figure 2.2). It depicts elements in the macro-level activity system from the first phase of the data collection cycle based on interviews with eight key individuals. It illustrates the *espoused* object of activity at the macro-level.

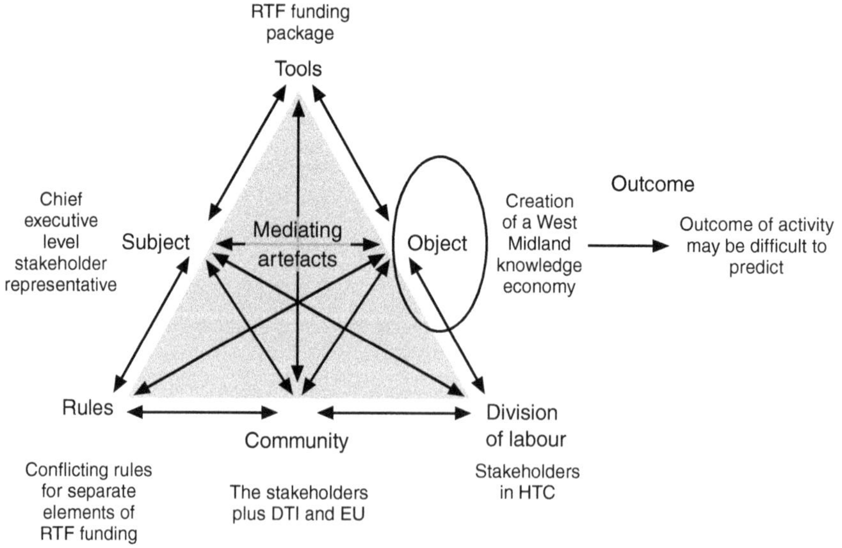

Figure 4.4 Macro-level object of HTC activity.

At meso- and micro-levels, this accounts for the emergence of variation in the *espoused object of activity* from that originating at the macro-level. This suggests that the whole *meta-activity system* shapes motivation and orientation towards an object of activity, and policy makers cannot predetermine this nor can it be imposed in a simplistic way.

In second phase, I interviewed fourteen members of the HTC steering groups. For these individuals, the object of their activity was orientated towards sub-regional priorities. The meso-level object of activity for each HTC steering group was orientated towards aligning the partners to address sub-regional problems.

Secondary data such as project plans and 'headlines' from the individual HTC Web sites were triangulated with interview data describing individual 'experiences' of the HTC policy implementation process.[1] Figure 4.5 shows how the data were mapped to elements of a *meta-activity system* at the meso-level and shows the *espoused* object of activity at this level of HTC policy implementation.

This set of interviews led me to people working at the micro-level of policy implementation who were responsible for delivering practical technology transfer projects to small and medium enterprises (SMEs). In third phase, my interviews focused on the practical strategies that projects adopted to make technology transfer happen for particular groups or clusters of companies with similar needs, interests, or dependence on specific markets. At this level, the communication between users and developers of new technology became really interesting. These data were triangulated with secondary data such as excerpts from reports from 1999 to 2004 analysing local problems. I also included published aims of HTC projects and materials from each HTC Web site.

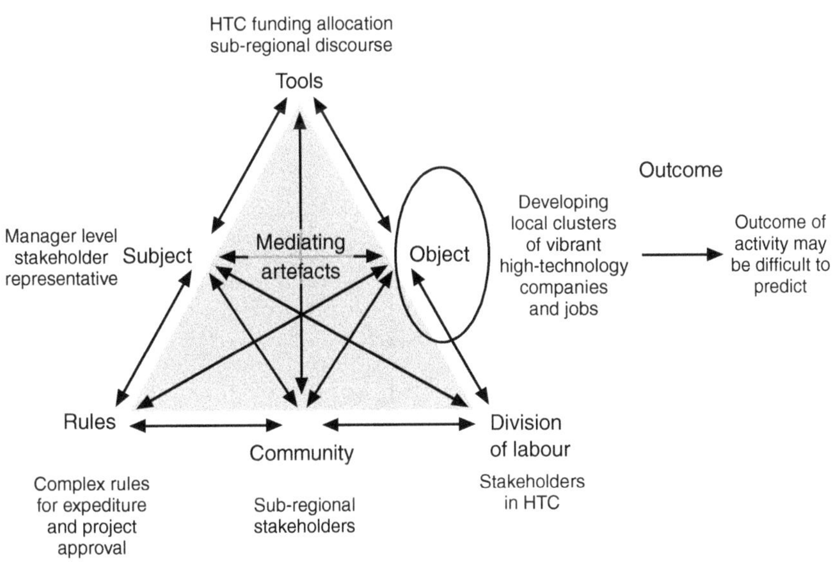

Figure 4.5 One of the meso-level objects of HTC activity.

During fourth phase, I pursued my interest in the *quality* of relationships between developers and users. Five more people were interviewed, and my questions pursued themes and gaps in data to reveal individual interests and organizational positioning of subjects representing stakeholder groups. I was able to combine this with data from third phase to construct an initial picture of activity at the micro-level. My interview questions were orientated towards capturing descriptions of collaborative practices, strategies, and tools that were used by HTC projects managers to help SMEs. Two detailed interviews with technical managers in SMEs provided rich descriptions of their perceptions of changes in their practices.

At this micro-level, the 'object of activity' was orientated towards transforming capabilities in the SMEs at the tertiary level in the Rover supply chain and to refocus the skills of their workforce on the development of new products. However, the regional policy may not have fully appreciated that complex innovations are difficult to implement (Fullan, 1999). These difficulties were increased because innovation was undertaken by newly formed multi-agency partnerships.

In particular, the notion of partnership was different at each of the three levels of policy enactment, and this had consequences for what could be achieved and explains why three characteristically different objects of activity emerged at each level. Figure 4.6 illustrates the inter-relationships between the elements of this *meta-activity system* and depicts the 'object' of activity at the micro-level.

4.2 Visualizing competing objects of activity

Objects of activity are multi-faceted and compete horizontally and vertically and over time. In this sense, 'objects' evolve in multiple dimensions. Expansive

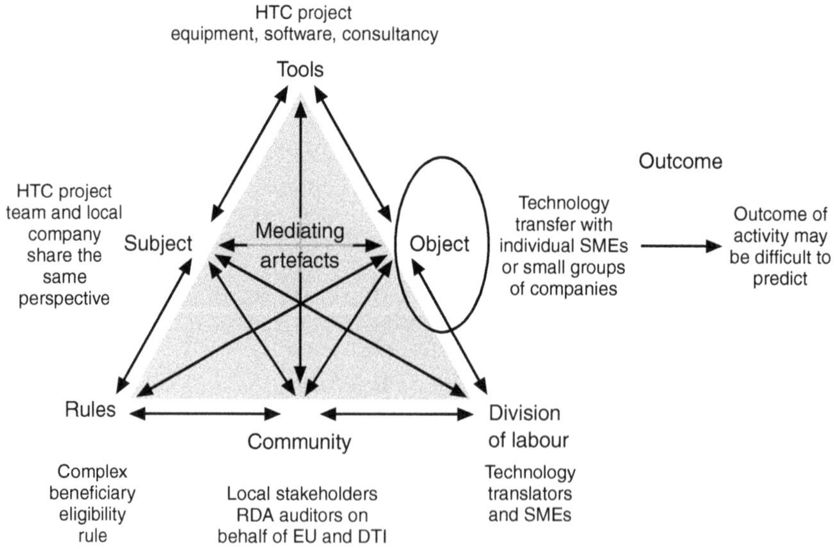

Figure 4.6 The micro-level: the object of activity in a project.

learning theory conceptualizes organizational learning as horizontal collaboration between one or more activity systems of relatively equal agency. This occurs in innovation where subjects such as individual experts make contributions of equivalent value, which tend to be complementary.

However, the gap in theory makes it difficult to depict relationships between organizations where there are differences of power. Currently, the theory does not offer an adequate explanation of what happens when activity systems with degrees of agency interact vertically. The term collaboration does not encapsulate interactions between activity systems where the power dynamics are imbalanced. Vertically linked activity systems in policy implementation are akin to the vertical relationships between manufacturers and the companies located at the secondary and tertiary tiers in the automotive supply chain.

Hierarchical levels of power and politics are embodied in these relationships, and the greatest degree of agency tends to be located at the top level. Figure 4.7 is adapted partly from the implementation staircase model and partly from a diagram used in a Scandinavian study depicting learning between production and management levels (Reynolds and Saunders, 1987; Toiviainen, 2003). It depicts the restriction on feedback from the lower levels caused by the hierarchical three-tier structure of HTC policy enactment.

The figure synthesizes the notion of the implementation staircase, and the original diagram designed by Toiviainen, and suggests that learning can flow in

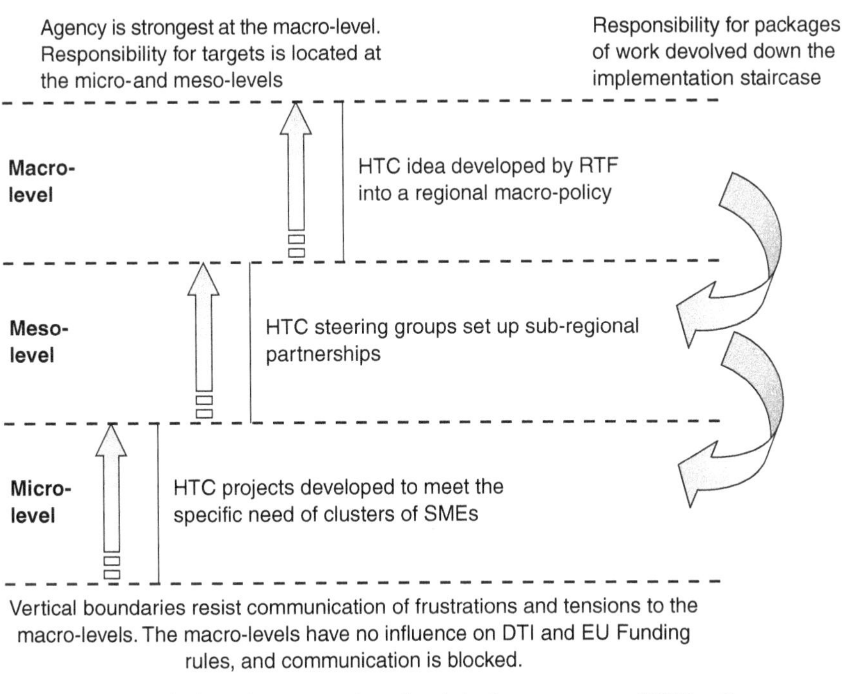

Figure 4.7 A depiction of agency at three levels in the enactment of HTC policy.

both directions (Reynolds and Saunders, 1987; Toiviainen, 2003). The figure illustrates the patterns from my data, where downstream communication flowed freely, but upstream communication was blocked. Toiviainen proposed that communication flowed equally well in both directions, but I found that the communication from the bottom to the upper levels was impeded by hierarchical relationships between strategic and operational actors.

Figure 4.7 depicts a tendency for those at the top level of HTC policy implementation to resist learning from other levels.[2] This is an interesting finding because expansive learning theory has not yet been developed to explain or depict competition between objects of activity of unequal agency. In other words, feedback from people at the operational or lower levels in the HTC implementation hierarchy tends to be disregarded.

For instance, the data captured interesting tensions exemplified in the comment of an HTC steering group member. As a member of an HTC steering group, the object of his or her activity related to sub-regional skill gaps. This was in tension with the regional perspective he or she adopted during his or her usual work role as a senior manager at the Learning Skills Council (LSC).

> I think that as a region we can't just rely on offering high technology, we need a broader West Midlands perspective that includes City A and Sub Region B. I think HTC2 is very territorial. To date I am not much involved with the other corridors – It feels a bit like a vacuum in terms of communication – I feel they are working in isolation – a bit like 'the people's republic of Anonymoushire', not looking at the West Midlands as a whole or considering the national picture. I think they will eventually miss opportunities because of this.[3]

This issue is discussed more fully in Chapter 6. The empirical data captured instances where there were problems in communication at the boundaries between macro-, meso-, and micro-layers. Although the levels did not seem to be hermetically sealed, the empirical data indicated that communication flowed downwards more easily than it flowed upwards. It showed that decision-makers at the macro-level were reluctant to be influenced by practitioners at the micro-level. However, members of the *meso-level* steering groups were aware that they could be instrumental in building consensus and linking perspectives of organizations that had no experience of aligning themselves towards achieving common goals and outcomes.

> In HTC2 it is important to understand the cross-sectoral agendas – and demystify the issues. I think it's about partners helping each other.[4]
>
> There was probably a gap between what the corridor wanted to do and what the corridor was.[5]

In this sense, members of the HTC steering group were aware of their potential to play a vital role in facilitating the flow of communication upstream and downstream.

4.3 Overview

Competing discourses from different subject perspectives shaped assumptions about the goal and purpose of working together. The next chapter explores the relationships and interactions across diverse work cultures, which are depicted as horizontal relationships between experts. Complementary skills can be combined when separate work cultures move closer together. Some projects were small scale, others were extremely ambitious. Where projects were the outcome of previous collaborations and the fruits of long-standing relationships, they were often successful. Some innovation projects are more accurately described not in the terminology of technology transfer but as the creation of a new artistic vision, which then results in a practical outcome, such as an innovative product.

Nevertheless, it was evident that many of the individuals and groups, on whom the success of multi-agency partnerships depended, had no previous experience of working together. They had previously existed in discrete worlds and had very little experience of collaboration on which to build. A series of *vignettes* illustrate the various ways in which boundaries between cultures were crossed. They include descriptions of the development of mediating tools called *boundary objects* to link the perspectives of two or more activity systems.

5 Horizontal expansion of the object by collaboration of experts

5.1 Horizontal expansion of the object of activity

From the perspective of expansive learning theory (ELT), horizontal expansion refers to the process of working on a shared problem with a group of people who see things differently. Interdisciplinary work creates tensions, which force everyone to reconsider their routine ways of working. It pushes people towards finding new ways of thinking and doing. While it is relatively simple to describe collaboration between people with complementary expertise, it is much more difficult to facilitate this in practice. Organizations of different types have particular cultural practices, and even companies who may be pursuing similar technological developments do not necessarily find it easy to collaborate. From this perspective, collaboration concerns the time and effort people put into developing positive working relationships.

It is important to recognize that because ELT grew from cultural and historical activity theory, it emphasizes the impact of rules, division of labour, communities and tools on the subject and object, and outcomes of activity. Fixed rules can be helpful. 'Even if collaboration proceeds gradually, it is important that participants establish effective governance mechanisms and have fixed rules regarding, for example, decision-making, sharing of responsibilities, and technical standards and quality maintenance capabilities' (JSBRI, 2005: p. 68). The Japanese Government has attempted to set out a rationale for companies to put time and effort into collaboration in technology transfer. First, they point out the benefits of fusion of different types of knowledge.

> In order for SMEs to suitably meet market demands that are changing and diversifying by the day, they engage in business in an integrated manner by collaborating with other companies with knowledge, technologies and know-how of different fields and effectively combining these resources, rather than relying solely on their conventional experience in their own fields.
>
> (JSBRI, 2005: p. 68)

Second, they point out the value of industrial partners collaborating with universities

> Partners for collaboration with SMEs lacking in areas of pure research and know-how regarding market development as described in the preceding section include a diverse variety of entities, such as middle-tier and large enterprises, central and local government research institutes, and universities, as well as SMEs in other fields.
>
> (JSBRI, 2005: p. 68)

Third, the role and contribution of each partner needs to be clear. In the terminology of ELT, this means being clear about divisions of labour. It also calls our attention to the risk of complicating the collaboration by including partners for the sake of it, regardless of whether or not they add value (as suggested in the notion of the tipping point).

> In order for collaboration to function effectively, it is important that there is a clear locus of responsibility for negotiations with customers, and that the costs of coordination should be cut by centralizing contact and coordinating functions. It is also essential to have an entity for overseeing and managing business activities as a whole, from product R&D and the manufacturing process to sales and the planning and provision of services, and liaising and coordinating between businesses. Unlike traditional subcontracting relations, such entities are core leaders that play a pivotal role in steadily developing successful collaboration.
>
> (JSBRI, 2005: p. 68)

This approach is both realistic and encouraging for companies who have no track record of collaborating with one another. Japan has a culture of rapid communication with face-to-face, telephone, and e-mails, all playing a role in keeping networks of colleagues up to date. In the EU, we risk imposing a rigid system of networking and clustering that attempts a superficial imitation of these arrangements. However, it may fail to capture the deep-rooted community spirited endeavour that underpins it and ultimately shapes its success. 'Building sustainable innovation networks can be perceived as a business cost or as a strategic competitive advantage' (Aoshima, 1993: p. 9). Despite the reputation of Japanese culture as more community orientated and more predisposed to sharing the costs and benefits of R&D across sectors and clusters, it is clear that the Government do not take 'collaboration' for granted. The Japanese Government put resources and effort into developing an appropriate structure of support.

5.2 How boundary objects bridge organizational cultures

Boundaries are referenced by many different theories of innovation and knowledge transfer. In ELT, the role of boundary objects in bridging diverse work cultures is highly developed. Although I have used ELT as a heuristic framework that explicitly refers to bringing two or more types of organizations together, it corresponds to the language of related theories that describe the same phenomenon as brokerage between communities of practice (COPs) or mediation of the distinctive perspectives of organizational partners. Expansive learning happens when diverse perspectives are connected and new practices are created.

In ELT, co-configuration is a shared creative process. To work together successfully on the shared creation of a new product, comfortable routine practices have to be abandoned, and this requires changes in the rules, the division of labour, and other tools of the old activity systems (Hasu, 2001: p. 15–17). The notion of mediation is at the heart of activity and of co-configuration.

Mechanisms called boundary objects help organizations to bridge conceptual and cultural gaps and smooth the process of innovation. In this sense, they are very important tools of innovation. Boundary objects vary in their nature and complexity and can include drawings, prototypes, and Web sites. Particular types of boundary objects can mediate learning across the activity systems in the partnership by supporting innovation in different ways.

From this perspective, it is possible to see how in the case of Japanese textiles, the 'definition of the problem' at the outset of a project acts as a boundary object. This amounts to a process of goal orientation, defining and agreeing the task in hand, without predetermining the outcome. Sharing the problem, and refining the terms and context of the problem, is a relatively simple matter where the partnership remains fairly small and tightly focused. In these circumstances, the object of activity can be agreed in a reasonably quick and straightforward way.

5.2.1 *Vignette: samples and prototypes as boundary objects*

The Japanese case study illustrates the advantage of focused small-scale partnerships. Usually Nuno is approached by a company with specific problems they wish to resolve by working in collaboration with Nuno. Designers were reluctant to use the term research and preferred to describe the process as experimentation by trial and error. Nevertheless, collaborative projects proceeded on a sound commercial footing, and there was a high tolerance of risk and creativity.

HB: What I'm interested in is how you communicate with textile manufacturing companies. Do you explain what you want from them with a prototype? Is this the kind of thing you use to explain a new technical process? Do you create something to show them?

KO: Yes – but it depends on how they approach us or how we approach them. We make something and many times we discuss this and then they make a sample. Always this is the first step. We spend time on this and continue the discussion. Sometimes we go there and sometimes they come here. If we

> have good communication with them and if they understand the idea I think it will always be a successful collaboration.
>
> (K. Otani, personal communication, 30 June 2006)

In communicating with small- and medium-sized manufacturers (SMMs) in textile industry, the design company used a *prototype* as a boundary object. This was a way to crystallize what they wanted the SMM to make, and it was effective as a two-way communication tool. In turn, the SMM created a *sample*, which was their boundary object to crystallize what they could manufacture to the designers. The interesting aspect of this use of boundary object is that an outcome from one activity system is treated as a tool by a different activity system. These boundary objects were instrumental in bringing the thinking of the manufacturers and the designers closer together. The use of visual models is common practice in engineering and the creative industries.[1] Whether a boundary object takes the form of a drawing, or concrete object, it can communicate an idea that only exists in someone's imagination.

5.2.2 Vignette: management consultancy project in the UK

HTC2 devised a project to provide customized multi-faceted business development advice and resource to small and vulnerable SMEs. A comprehensive package of business diagnostic tools and practical support was designed to cross over the boundaries of expertise to reduce the challenges faced by the company. A holistic analysis of needs aimed to reveal 'in-house' weaknesses in knowledge and experience. 'It might help with finances or ... market research or reviewing the management structure of the company ... to fund Field and Beta testing, or to develop prototype and get through the proof of concept stage, prior to full commercialisation'.[2] The project customized for each SME an integrated support package, which included financial assistance, expert advice, and practical help from experts in technical aspects of production or marketing. Using the terminology of ELT, it brought together discrete communities of experts to work towards a shared object in collaboration with each company. This approach set aside the usual divisions of labour to focus concerted effort from different experts on the needs of each individual SME. Unusually, this was related to a suite of complementary projects in the HTC, and it was therefore well connected to the meso-level, although not to the macro-strategic level.

5.2.3 Vignette: projects connecting industry with art schools

There is a clear cross-fertilization of thinking between the UK and Japan on this issue. It is an example of policy borrowing on a micro-scale. The practice of industry sponsoring student projects in the UK seems to have influenced the way in which the company nurtures new designers when they undertake their first few collaborative projects with industrial partners.

> I studied in England and I found that the companies and the colleges are closer. I thought they worked together more than in Japan. In Japan the art college and the industry are completely separate and different from one another. In England many companies invite art students to collaborate with them. I mean the sponsorship in England where students work on a design with a company. In England the system brings them closer. In Japan we don't have this system. The college and the company are separate.
>
> (K. Otani, personal communication, 30 June 2006)

This design-led innovation has had an international influence on the perception of textile artists as commercially successful innovators. Designers from Nuno Corporation initiate creative collaborations with industrial partners that continue to inspire art and textile students around the world.

5.2.4 Vignette: collaborative projects to develop electronic tools

The UK case studies illustrate various ways that material artefacts and 'conceptual models' act as boundary objects. The next two vignettes describe virtual or electronic tools that were developed in a collaborative process and also acted as tools for further collaboration. After the initial formation of the HTC steering groups, attitudes among individual members of the HTC steering group began to change rapidly. Initial suspicions and tendencies towards self-interest evaporated as confidence in the evolving relationships grew. Along the way, the HTC1 steering group decided to create a Web site. This acted as a *boundary object*, which helped to forge allegiance and a shared identity for their members. 'Where I would have been fighting for funding for Phase 3 of the Malvern Science Park – we have together (the HTC1 Board) put in an action-planning bid with better leverage with AWM and GOWM'.[3] The steering group decided to create an electronic database as a resource for a shared website. The process of making the database connected perspectives and the artefact itself, and it was used as a communication tool by the partner organizations, so the project acted as a *boundary object* in two ways. The project was instrumental in building trust, which then forms the basis of confidence in accepting more difficult challenges involving greater degrees of risk. It changed 'organizational behaviour', and people became more open to innovation with each positive experience of multi-agency learning.

> The corridor has learnt the value of making partners aware of all that is going on in the region, and of connecting things up by facilitating electronic communications between the partners at steering group and project level. In particular an electronic tool to link the databases of the 'partner organizations' has been effective. This tool can find out which companies have been helped by the various schemes and identify gaps in provision and access.[4]

The database on the website became a valuable tool to bridge the boundaries between organizational cultures. In the process of creating the boundary object and using it as a tool, horizontal collaboration or polycontextuality at the meso-level of HTC policy implementation grew.

Although there had been no involvement of people from other policy enactment levels in the creation of the database on the website, it could be used by anyone, but it was the commitment to the task and the understanding that resulted from working together that made the 'cultural change' towards expansive learning. In other words, the boundary object creation process was more significant than the database artefact created.

Figure 5.1 illustrates how a Web site acts as a tool to connect perspectives. In other words, the Web site was an especially useful boundary object because 'the process of making it' connected different activity systems by initiating the active cooperation of members of the steering group.[5]

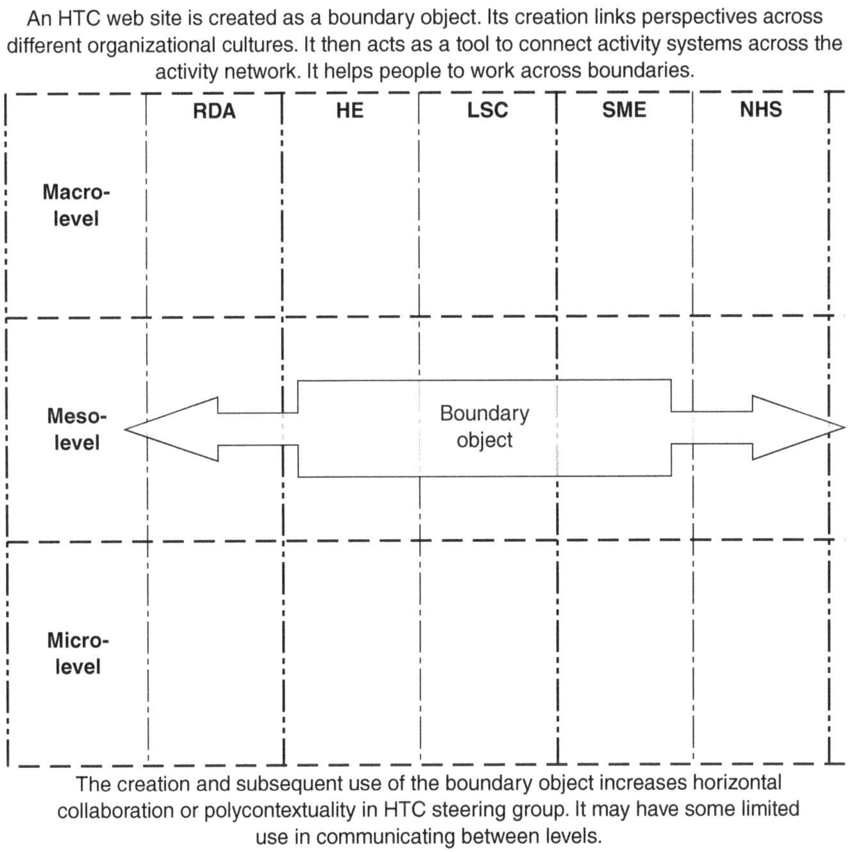

Figure 5.1 The creation of electronic tools as boundary objects.

5.3 Users and developers: consumer-oriented innovation

Mass customization is a term that has become widely used to describe the change in thinking from mass production to a more consumer-orientated approach by product developers. Marketing experts are constantly updating their research on consumer preferences and purchasing trends. Nevertheless, there is an assumption that this trend is most developed in the USA and Japan.

HB: In the US is there closer involvement of consumers within product development?

DA: Not as much as the myth would have you believe! There is still reluctance – so I often say 'when all else fails ask a customer'. Venture Capitalists usually encourage companies to talk to customers or their potential customers very early in the development of their idea or product.

(D. Aslin, interview transcript, 11 October 2006)

The technologies to enable shorter product development cycles have coincided with increased consumer demand for individualized products and services. Designers are very aware of the needs and aspirations of the consumer and can ensure that consumer preferences are built into the process of new product development.

5.3.1 Vignette: interacting with clients in the design process

Since 1945, documented trends in Japanese textile technology show that the emergence of a world-class textile technology in Japan resulted from close cooperation. Competitiveness and effective innovation required collaboration between all segments of the industry, including producers of fibres, yarns, fabrics, apparel, and processing equipment. While productivity (output per machine-hour) in yarn manufacturing and fabric formation increased several-fold over the last two decades, the speed of finishing improved only moderately. Consequently, advances were sought in dyeing, printing, heat transfer, chemical application and recovery techniques, and fabric stabilization.

> The much fiercer pace of change in clothing fashions compared with other finished products creates particular difficulties, making it hard to predict demand each season. To date, ascertaining demand for products has often been left to wholesalers. In the future, however, diversifying tastes and shortening product cycles will make it increasingly important to accurately monitor customer needs through face-to-face communication.
>
> (JSBRI, 2005)

The Japanese Government's 2005 White Paper reported, 'There is a decline in the position of Osaka's textile wholesalers, and the relocation of head offices by apparel manufacturers, sales to other regions, and in particular to the area around Tokyo' (JSBRI, 2005). The recognition of inevitable changes in the mechanics of the market for textiles is evident in policy documents and advises SMEs and SMMs of the importance of close communication and collaboration with consumers. 'Improved manual garment assembly and progressive automation will provide the flexibility needed for small-lot and even one-of-a-kind production' (Berkowitch, 1996). Innovative designers began to experiment with combinations of man-made and natural fibres, and as a result, a vibrant market for new kinds of textiles was created. Japan led the world in the development of shape memory fabrics and in unusual combinations of fibres, with properties suited to niche markets such as sports or the catwalk. Personalized clothing adjustable in temperature to maximize comfort and clothing woven with fibre optics presented new opportunities for creative design. Nuno is a good example of a Tokyo-based design company that established an international reputation for leadership in this area of work. The collaborative practices of designers are orientated towards aesthetic and creative goals, and their object is creative rather than technological. Several vignettes from the case study illustrate that designers would not describe the 'object of their activity' as innovation or knowledge transfer. They prefer to use the term 'experimentation'. Technology itself is seen as a tool to enable new creative practice. This practice involves the creation of new products such as new textiles that include unusual combinations of metals and new types of synthetic fibres.

> Nuno values face-to-face interaction with customers in our shops. We designers serve the customers and in doing so get valuable feedback on what features the customer finds most attractive. Feedback from the consumer or user is also helpful in indicating new directions for design ideas. In this sense developer – user relationships flow naturally from the way in which Nuno Corporation conducts business. Customers are able to communicate directly to Nuno designers their preferences in colour, form, and texture of fabrics.
>
> (K. Otani, personal communication, 30 June 2006)

It is motivating for the designers to get immediate feedback on products, and also, the customers feel valued and relish the chance to talk with creative individuals who are keen to listen to their views and integrate customer feedback into future products.

> Yes we always have graduates wanting to work with us. Sometimes if we have a job they join us and start from the beginning.... The people who work here have to do all aspects of the work not just design, looking after the shop, interacting with customers, marketing, administration and so on.
>
> I learned here. We are always learning. I still learn everyday. I have my own projects and I have responsibility for everything in the project. Often I involve other colleagues.... We have good teams and always discuss the

> project and share our experience. Nuno is quite unique and different from other companies in Japan I think. I don't know so much about other companies, but I think sometimes they have a marketing person and a design person whereas at Nuno we share our information and learn to do everything.
>
> We don't design everyday. Just before a collection we work on our designs. Sometimes we work on the collection together and everybody employed in Nuno is a designer and does designs for the show. Sometimes we work here together and sometimes we work at home. Reiko is the chief designer and sets the direction for everything. She checks all the designs and we discuss and the colour and size and so. She decides which designs are suitable for weave or print and on which fabric. Reiko also decides which designs will be most attractive overseas, and should go to the Australian or US market.
>
> (K. Otani, personal communication, 30 June 2006)

5.3.2 Vignette: Nuno collaborations with fashion designers

The success of new fabrics is dependent on responding to markets with fast changing and diverse customer needs (Aoshima, 1993: p. 9). For textile designers, the fashion designers are consumers of innovative products. In this sense, one industry is related to others in a supply chain. Collaborations to achieve particular effects depend on integrating customer feedback into the development of new fabrics.

Nuno's textiles are sought out by leading international fashion designers for their impact on the catwalk. Similarly, architects are keen to collaborate with Nuno designers to create special effects with textiles that change the aesthetic and emotional ambiance in the interior spaces of modern buildings (Millar, 2005: p. 8).

Consequently, Nuno designers have long-term relationships with a network of textile manufacturing companies (SMMs). This gives Nuno competitive advantage through rapid product development and the timely introduction of sequences of new products. 'We, at Nuno, always design our textiles in-house. However, for individual projects, we work very closely with other creators, such as fashion designers, costume designers and architects' (R. Sudo, personal communication, September 2006). A new class of organic materials, sometimes referred to as 'smart' because of their ability to respond significantly and reversibly to environmental changes, have been commercialized in Japan with the promise of revolutionizing a great many articles ranging from fibres and textile finishes to mouldings. The focus of shape memory polymers is based on a family of polyurethane block co-polymers, made of alternating soft and hard segments, which show large changes in their physical and mechanical properties at their glass transition temperatures.

> Advances in conventional and smart materials and human sensory perceptions coupled with a one-of-a-kind production capability could usher in

> a time when the industry might literally personalize apparel to meet the physiological and psychological comfort requirements of each consumer.
>
> (Berkowitch, 1996)

The decline of Japanese traditional local industries, and the falling demand for kimonos coupled with the increasing import of inexpensive textiles from neighbouring countries, led some companies to despair.

> Actually there are many manufacturers who market their own creative textiles. Originally, the Japanese textile industry was made up of many separate professions. For example, the textile distributor would have a design room that would send original designs to each factory. What the factories, or manufacturers would produce was totally controlled by the textile distributor. As their business was solely on a sub-contractor basis, manufacturers or weavers never had any design rooms in the past. All design ideas would come to them from the textile designers at the apparel and interior companies they did business with.
>
> (R. Sudo, personal communication, September 2006)

Increased competition motivated such companies to innovate, while preserving traditional skills, and many approached Nuno for help (Millar, 2005: p. 10). 'We are not manufacturers – we work with several industrial manufacturers all over Japan. We are a textile design company, which includes producing original textiles. This is why many people mistake Nuno for a textile manufacturer' (R. Sudo, personal communication, September 2006).

5.3.3 Vignette: public–private sector partnership zero-energy homes

The Sacramento Municipal Utilities District (SMUD) might seem an unlikely organization to commission research, but it does just that. When there is a gap in the market and no product to meet a consumer demand, they find a company or a university willing to work with them on the problem.

> What works for us is an approach where we form programmes around the issues that our customers have prioritised. We recognize that we are not manufacturers – but we can try to set up partnerships that are about making products e.g. the Zero Energy Homes. We are close to the market and we can tell manufacturing partners about those needs and even pay for some of the development. It helps us both. The approach can also attract State of California funding because they see the state-wide benefits.
>
> For instance we found that over 80% of the big house building companies used concrete roof tiles – and no available photovoltaic products could integrate with them. So we funded the R&D in two different manufacturing companies to create a photovoltaic tile that could be used with these concrete tiles. Now these are very widely used not just in Sacramento, but also in most other areas of California.

In this sense, SMUD acts as a champion of the small environmentally conscious consumer and intervenes where there is a market failure. At the same time, they stimulate broader interest in sustainable energy by making it a truly viable option.

5.3.4 Vignette: collaboration between users and developers of photonics

A leading company in the Hamamatsu Photonics cluster based in the Shizuoka region of Japan agreed that innovation happens when different organizations and people work together. 'We think it is also applicable in Japan, that "when world's collide new insights arise", although we cannot think of any specific examples to give right now' (Photonics Hamamatsu, personal communication, May 2006). The company is keen to involve consumers in the process of innovation. The company makes special low-voltage lighting devices for use for diverse industries including automotive and medical technology manufacturers. Their consumers are therefore not individual members of the public but other business people who can use their photonics devices in many different types of products. 'As the development of new technology is aimed more toward producer-driven solutions than from the opinions of end users, we rather value development from our own (original) technology' (Photonics Hamamatsu, personal communication, May 2006). The photonics company involves its consumers or end-users in the development of new technologies by asking for feedback on prototypes of new products. In this way, they are able to get a sense of the likely enthusiasm and consumer demand before they move on to full-scale production. 'Although our customers are not general consumers but manufacturers (makers), the opinions of end users are highly instructive and informative in developing innovative products, or in innovating "product development"' (Photonics Hamamatsu, personal communication, May 2006). In other words, if a customer expresses a need for a product that does not yet exist, it would motivate the photonics company to carry out research and development activity. In this sense, this would lead to experimentation to test whether the need expressed by the consumer could be transformed into a viable commercial product.

5.3.5 Vignette: how users transformed an outcome into a tool

In particular, vignettes from the UK case studies show how boundary objects are instrumental in collaboration and bridge the distinctive perspectives of the users and developers in the process of innovation.

The policy rhetoric speaks of the demand side for technology transfer in a rather simplistic way, which assumed that demand would be consistent and homogenous. However, experience suggests that depending on the size and type of companies, they would want different things from research. For this reason, technology *pull* from users was unlikely to be in the same direction. The empirical data showed that the 'demand' for innovation and technology transfer was extremely inconsistent and varied. Users wanted a diverse range of advice and practical support from 'developers'.

At the micro-policy implementation level, user demands were articulated locally in negotiation with the technology developers. In practice, demand from groups of SMEs had resulted in the creation of HTC projects to match their specific needs for new technology and advice on diversification to new markets.

Given that industry investment in research has been lamentably low in recent years, the capacity to recognize the value of new technologies to local companies seemed questionable (Brown, 2004a: p. 1).[6] In the absence of a strong commitment to research by local companies, the public sector was 'left to drive the technology transfer agenda'. The task fell to the universities, the Regional Development Agency (RDA), and QinetiQ.[7]

Connecting local initiatives to achieve complementary effects was done at project level, and this helped to engage the enthusiasm of SMEs. For instance, an HTC project helped polymer manufacturers to develop new products and markets, because demand from the automotive industry was declining rapidly. Another HTC project assisted small engineering companies by bringing new technologies and processes into the local SMEs, at the tertiary level of the Rover supply chain.

In the process of 'new product development', the developer perceives the artefact as the *object of activity*, whereas the user sees it as a *tool*. Bridging communication between developers and users involves linking different perspectives. Working across boundaries between diverse activity systems can be challenging because there are few precedents and little recognition of the problems encountered. Interaction between actors and social groups in the HTC involved working in a 'contested terrain' because 'technology transfer' is the domain of neither research nor commerce but a *hybrid* of both. Setting up HTC projects created a context for interaction, which encouraged fluidity in identities and social interactions.

Individuals who are members of both communities understand the perspective of both worlds and can act as translators. They can make sense of things and explain in the language and conceptual vocabulary most easily understood by scientists or by industry. In this interface, social relations are characterized by insecurity as a consequence of multiple identities and strategies (Ludvigsen *et al.*, 2003: p. 307).

However, in the HTC projects, the key individuals who acted as 'technology translators' interpreted a new technology so that a small local company could understand how it could be applied in a commercial production process. This led to expansion of the activity network so that the users become involved in the co-construction of a new expanded object of activity. Some HTCs devised projects to deliberately involve users and developers in co-construction of new products.

> Developers and users do not work together particularly well at the moment. It is probably a challenge for the future. We're trying to take a holistic view; we talk to companies about clustering and about how Universities can help them to tackle technology transfer.[8]

In one HTC project, a great deal of time and effort was invested in gaining a better understanding of co-configuration.

> A working group is considering the **push** (by Universities) versus the **pull** (exemplified by the Business Links on behalf of the local companies). We don't yet have the structures to understand how this process could work better. We know that there is SME demand and SME need. We are interested in how practitioners see this and we want to understand practices that would make this work better [emphasis in original].[9]

Each HTC devised different strategies and specific projects to capture the interests of local SMEs and to make the HTC concept relevant to them. The HTC project teams realized that they would need some tools such as rapid prototyping equipment to motivate SMEs to change their behaviour. 'You are not very interested in modernising or diversifying until external pressure makes that a necessity. Engaging particularly with the SMEs in this process has been very problematic'.[10]

5.4 Overview

This chapter showed how complementary skills can be combined when separate work cultures move closer together. It depicted the relationships and interactions across diverse cultures as 'horizontal relationships between experts'. Collaboration was difficult when the individuals and groups had no experience of working together. A series of *vignettes* illustrated the various ways in which boundaries between cultures were crossed. They included descriptions of the development of mediating tools called *boundary objects* to link the perspectives of two or more activity systems.

I used vignettes as a presentational device to illustrate the complex theoretical implications of unfolding activities as they occurred in a specific location at a particular moment in time. They sketched the different ways that material artefacts and 'conceptual models' acted as boundary objects. The vignettes described instances of an innovation project being an outcome of activity, which was then used as a boundary object. In this form, it became a mechanism to align the partners around a larger common goal and more sustained endeavour as illustrated by the Web site described in vignette 'collaborative projects to develop electronic tools'.

I found that small collaborations inspired confidence and a sound footing for more extensive collaboration. For instance, vignette 'public–private sector partnership zero-energy homes' showed that if partners collaborate in the development of a boundary object such as a prototype solar roof tile, it is an outcome which may also be used as a communication tool. This can help researchers to work alongside the private sector and collaborate in the commercialization phase.

The next chapter describes the impact of boundaries between levels. Boundaries between levels are less permeable than those separating different cultures described as horizontal relationships in this chapter. Strong boundaries make expansive learning between levels more complicated and difficult. Chapter 6 focuses on the role of individuals, who can act as *boundary mediators* between relatively powerful and weak activity systems.

6 Mediation between activity systems of unequal agency

6.1 The significance of power and hierarchical structures

Many theories of innovation do not address the issue of power. Community of practice theory (COPT) and actor-network theory (ANT) are somewhat disappointing, and even expansive learning theory (ELT) sidesteps an analysis of power dynamics in organizational partnerships. ELT is rather good at explaining collaboration between experts, but current theory does not explain collaboration between organizational partnerships characterized by political influence and power struggles.

Although ELT offers a convincing account of the ways in which boundary objects can help experts to communicate along the horizontal plane, organizational partnerships also need to be connected vertically to link communications between people in organizational hierarchies. It does not explain how unequal partners become oriented towards a common object of activity.

Given the conceptual borrowing and cross-fertilization between activity theory and COPT, it is not surprising that Wenger also uses the idea of an individual who moves between different communities of practice. In COPT, this role is embodied in the concept of a 'broker'.[1] A 'broker' can act as a bridge between boundaries that separate communities so that 'new knowledge can then be applied and expanded in new projects, and so the cycle goes on' (Wenger, 2000: p. 237).

Boundary crossing is also an important conceptual and practical feature of complexity theory, which has influenced the triple helix theory of innovation. There is agreement between the perspectives of complexity theory, ELT, and COPT that there are fuzzy boundaries between the phenomena and organizations (Leydesdorff, 2005). The triple helix is characterized by complex dynamics, where each system remains in flux and reconstructions evolve continuously. Since the triple helix systems of universities, government, and industry are increasingly knowledge intensive, their borders remain uncertain (Leydesdorff, 2005). It is assumed that through this interaction all three types of organizations learn and are enabled to work together more effectively. It is assumed that this will happen regardless of the relative impact of power of the partners in terms of authority or money. Regions with research universities could expect to experience a growing demand for knowledge transfer to industry and, through government, to society; however, this effect cannot be taken for granted.

The triple helix model has been shown to work poorly in non-metropolitan regions in countries such as Australia (Cooke and Leydesdorff, 2006). In general, industry has been less willing to pay more for privileged access to knowledge-based growth opportunities by funding research, stimulating closer interactions among the three institutional partners than it was assumed. Consequently, the growth of incubators and science parks and stimulation of academic entrepreneurship skills has been more difficult to achieve in practice than policy or theory would suggest.

To summarize a point, 'horizontal complexity' is about working across systems or sectors and agencies. 'Vertical complexity' relates to changes at individual, local, and regional community levels and the interactions between these levels (Connell *et al.*, 1995: p. 2). In applying ELT in two dimensions, I risk working at the edge of an evolving theory and in uncharted territory. This chapter concentrates on crossing boundaries between levels where the activity systems had asymmetrical status and agency. Depicting the complex social, political, economic, and physical dynamics simultaneously and in both dimensions was extremely complicated. It is a problem I have been unable to solve in my research nor is it adequately resolved in previously cited studies of innovation (Hasu, 2001; Toiviainen, 2003).

6.2 Vertical expansion of the object of activity

Innovation projects tend to be characterized as the horizontal collaboration of experts; however, insufficient attention has been focused on opportunities for vertical interactions between levels. The motivation to create boundary objects to connect activity systems is weak if one of the 'partner organizations' is significantly more powerful than others. Vignettes from the case studies illustrate some of the ways in which 'partnerships' can take active steps to expand the object of activity vertically to connect activity systems that differ in dominance and power. Unless steps are taken to connect the perspectives of dominant and weaker partners, the capacity of organizational partnerships to learn will not be fully realized. This is illustrated in Figure 6.1, which shows the impact of power on the flow of communication in high-technology corridors (HTCs). It is adapted from a diagram of the implementation staircase, designed in 1987, to illustrate the transformation of national education policy by teachers working in schools (Reynolds and Saunders, 1987). My data captured a more complex pattern. Figure 6.1 shows how people in powerful strategic positions blocked 'feedback' from actors at the operational level.

Vertical boundaries seemed to be experienced differently by various people attempting to move between levels. The data showed that key individuals were able to act as *boundary mediators* only if they belonged to the strategic upper level. If operational actors attempted mediation, they were blocked by the boundaries separating levels. A hierarchical implementation structure could be bridged

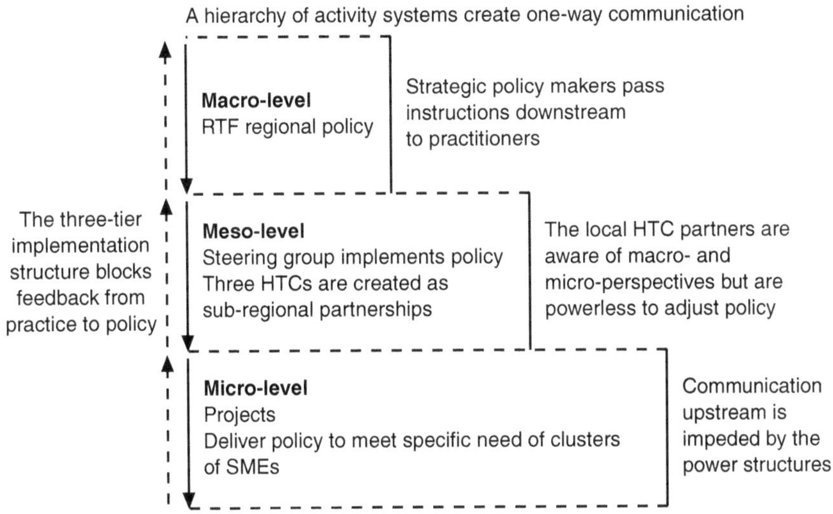

Figure 6.1 Communication between powerful and weak activity systems.

by someone influential acting as a boundary mediator. They connected the perspectives of practitioners delivering technology transfer projects with the people responsible for strategic planning at the macro-level. Although a boundary mediator is a member of multiple communities of practice or activity systems, the mediator retains a coherent sense of self-identity. Engeström describes mediators as 'boundary crossers', and they can resolve tensions and conflicts that prevent the partner organizations working in harmony.

> Boundary crossers are people who move from one activity system to another and work in two or more activity systems simultaneously, 'sharing the boundary object and the work based on it' with some partners in the other activity system.
>
> (Tuomi-Grohn, 2003: p. 203)

Boundary mediators can bring together weak and powerful activity systems. Figure 6.2 illustrates how the special status enjoyed by boundary mediators allows them to travel through boundaries in either direction, connecting perspectives along the way. They have a unique capability to bring to the attention of the strategic level issues from practitioners at the operational level.

The data from the HTCs showed that a 'boundary mediator' can range freely between levels and can be considered to be fluent in the languages of different activity systems because they understand and 'are understood by' all the various communities. The next section features vignettes, which illustrate the link between the style of collaboration in a partnership and the distribution of power among the partners.

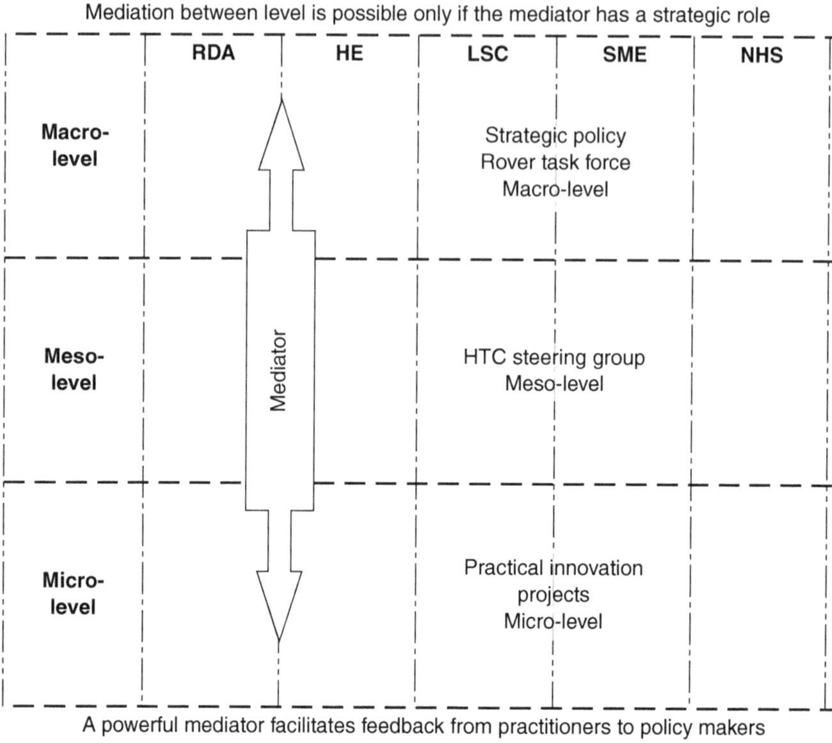

Figure 6.2 A depiction of the vertical levels bridged by boundary mediators.

6.3 Collaborative projects as cycles of expansive learning

Technology transfer projects can act as catalysts for organizational change. Projects can initiate changes in practices, which include modifications to established routines and behaviour. The experience of collaborating with people who take a different perspective drives organizations to reflect on 'what they do and why it's done in that way'. Innovation or 'knowledge transfer' projects create opportunities for people to re-align relationships between weak and strong partners. The power dynamics can be visualized as tensions between vertically arrayed activity systems. In practical terms, 'projects' can connect organizations of unequal agency by creating conceptual bridges between vertically arrayed activity systems. It is possible to visualize each project as a self-contained cycle of expansive learning, and yet at the same time, a project can build trust and confidence for more extensive collaborations leading to more ambitious expansive cycles. Projects focus on practical tasks and create opportunities for partners to work together towards an outcome. The process of collaboration creates 'neutral spaces' where it is possible to see the problem or challenge from another organizational perspective as if with 'new eyes'. Such 'neutral spaces' are conceptually close to the collective zone of proximal development, which is described in ELT as 'the gap between old and new organizational routines or practices'. If the gap is too large, it is too difficult to make a change. On the contrary, if the gap is too small, the change will be insignificant and will not move the organization forwards.

6.3.1 Vignette: Nuno's sustained investment in collaboration

Nuno Corporation makes a long-term investment in building relationship with clients, including individual consumers who visit Nuno's shops and exhibitions in the USA and Europe to buy textiles. Nuno also nurtures relationships with experts from other creative disciplines. Reiko Sudo encourages Nuno designers to develop relationships with other professional communities of practice. This has been highly successful and consequently architects and fashion designers collaborate with Nuno designers to achieve a particular visual, aesthetic and emotional impact.

Nuno received a request from Mikawa, which is one of the most ancient weaving areas in Japan with local textile records going back to AD 799. Although some textile manufacturers of Mikawa were engaged in innovating with pleating and polyester textiles, many other manufactures were frustrated because they simply received orders with no opportunity for creative input. One such company had reached a crisis point when they approached Nuno and asked for guidance in adaptation of heritage techniques to meet modern needs.

> It wasn't really our normal area of practice, but in order to work out what they could do, I had to really understand what they'd done in the past, and what they still could do. We ultimately settled on adapting their time-honoured patterns to create a portfolio of new fabrics around the theme of traditional Japanese toys. Reiko Sudo quoted in.
>
> (Millar, 2005: p. 22)

A project of this type is also valuable as an investment in long-term relationships. In other words, experience of collaboration builds trust and open communication, and these are not achieved instantly nor can they be taken for granted.

> I see great potential in their intelligent fibres, polyesters that contain special particles that dissolve in alkali, leaving lustrous surface textures of pinprick holes. I'm sure Nuno will be doing more with this in the future'. Reiko Sudo quoted in.
>
> (Millar, 2005: p. 22)

Collaborative innovation projects can help to transform the thinking of small vulnerable companies. For instance, a Nuno designer described similar challenges facing textile manufacturers in UK and Japan:

> Yes, we have the same problem – in some industries many small companies are closing down because the next generation does not want to continue in

> that business. A lot of businesses like printers are closing down. We try to work with them. Often young fashion designers or designers at Nuno try to work with them. It can help us both. It is difficult because there are some big changes in the industry, and production in China and India is very cheap.
>
> (K. Otani, personal communication, 30 June 2006)

In summary, there is a huge risk that collaborative development projects are treated as discrete, short-term activities orientated towards a specific product development. However, this misses huge opportunities that arise from a more sustained approach.

> However, in practice there are direct relationships among development projects, especially where new products are repeatedly introduced into markets. In such cases, improving product development capabilities, and inter-project learning, through the sequencing of multiple development projects may be as important to successful product development as developing advanced component technologies.
>
> (Aoshima, 1993: pp. 16–17)

Nuno makes a substantial investment in nurturing sustainable relationships. Nuno takes the intention of innovation seriously and makes an emotional commitment to the 'object', although they may describe this in terms of a creative objective rather than as a corporate goal (Hellström, 2004: p. 647). Nuno's commitment is the reverse of the so-called demand-push and supply-pull explanations of innovation (Hellström, 2004: p. 647). In a sense, Nuno tends to create numerous small experimental projects, which can act as boundary objects and catalysts for more extensive industrial change. This method enables customers to be involved in the development of new products in an uncomplicated and effective way. This sustained commitment stands in sharp contrast to the approach adopted in much EU-sponsored innovation, which treats projects as isolated and discontinuous experiments.

6.3.2 Vignette: Hamamatsu Photonics cluster

The Japanese Government offers support to many key industrial sectors, and the Ministry of Economy, Trade and Industry (METI) promotes a realistic attitude to the costs and investment in collaboration as a sophisticated means of responding to market needs. Japanese policy continues to be strongly influenced by theories and practices developed in the USA.

> In order to support the independent development of regional industries, numerous regions are also pursuing 'cluster plans' designed mainly to develop ties between the individual enterprises, universities and other entities that form regional communities, rather than focusing on physical regions, as has been the case to date. By such means, it is hoped that the

> formation of regional human networks will be encouraged and the conditions established for the creation of innovation, leading in turn to regional economic revitalization.
>
> (JSBRI, 2005: p. 124)

Regional policy encouraged the formation of a photonics cluster in the Shizuoka district. One significant company working in this cluster explained that recent collaborations with academia were subsidized by the Japanese Government under the basic plan of science and technology.

> In recent years we have actively participated in over ten projects with differing themes every year, in various contracted and subsidised research projects, such as the New Energy and Industrial Technology Development Organization (NEDO) under the Ministry of Economy, Trade and Industry and JST under the Ministry of Education, Culture, Sports Science and Technology.
>
> (Photonics Hamamatsu, personal communication, May 2006)

When the initial research plan of a university and the technical needs of a private sector enterprise match, the private sector enterprise (sponsor) provides research funds to a technology licensing organization (TLO). NEDO then grants the TLO a subsidy of up to double the amount of the funds provided by the sponsor. Thus, the programme promotes the creation of new industry and markets by utilizing technologies developed by universities. It is possible to apply for a grant covering two-thirds of the total budget (up to the double the funds provided by sponsors) and up to ten million yen (NEDO, 2006).

It is often reported that in Japan, there are excellent communications within and between companies with common needs. It is also suggested that the 'sense of community' may be stronger than other countries.

> The Japanese are less individualistic, more community spirited and have more patience than people in other countries/cultures. Alumni links with Professors ensure companies network with academia in pursuit of large technology transfer goals. Research, Development and Innovation are more strongly linked and the application of research is highly valued and engineering is a highly respected profession.
>
> (Eagar, 1986)

The Japanese Government has research evidence that shows that enterprises engaging in 'collaboration with universities and research institutes' to develop more advanced technologies and in 'collaboration with customers' tied directly to market needs were more likely to have rising ordinary profits than enterprises that innovated independently (JSBRI, 2005: p. 67).

> Time and money must be spent on nurturing trust, and even if a suitable (commercialising partner) can be found, technical evaluation in the 'fusion'

> process is tricky – and adjustments must be made between the two sides. There is therefore no incentive to engage in collaboration unless there are advantages to be gained, such as an increase in value added or reduction of costs.
>
> (JSBRI, 2005: p. 66)

The photonics case study characterizes this type of METI policy. The photonics cluster has created opportunities for collaborating with companies with similar interests and needs. The NEDO was established by the Japanese Government in 1980 to develop new energy technologies to reduce oil consumption. Eight years later, in 1988, NEDO's activities were expanded to include industrial technology research and development and, in 1990, environmental technology research and development.

> Although the number of cooperative activities is quite small, compared with companies in other industries, we do have some opportunities for collaborative research and development with other companies in the same industry. Collaborative research work with other companies in the same industry involves some difficult issues, such as dealing with intellectual property rights. However, we consider that collaborative research and development with other companies in different industries (such as companies which are potentially our users in future) enables us to grasp the needs of the end users and to enjoy a significant advantage in setting an appropriate direction for the development.
>
> (Photonics Hamamatsu, personal communication, May 2006)

Photonics is one of the energy-saving technologies where NEDO grants can bring university expertise to support clusters of small companies. The Hamamatsu Photon Valley Initiative also funds activities under the scheme run by the Japan External Trade Organization (JETRO).

> The government (Hamamatsu City Office), a research institute (Organization for Hamamatsu Technopolis), an industrial association (Hamamatsu Chamber of Commerce and Industry), and some of the more aware medium-sized firms are now promoting clustering around optical science as the core.
>
> (JETRO, 2004: p. 43)

Hamamatsu was also awarded funding to implement the Government's cluster-related projects such as the Intellectual Cluster Creation Project and the Collaboration of Regional Entities for the Advancement of Technological Excellence (CREATE) in the region. This aims to promote closer communication and new collaborative ventures with overseas investment.

> The actors for regional cluster development, however, consider that the creation of a new cluster centred on optical science will require introduction of

> the dynamism of foreign firms into the region. They are now disseminating information to foreign firms and preparing the environment for attracting them.
> (JETRO, 2004: p. 43)

There is a growing awareness that in the field of optical science, it is important to establish a strong presence in the global market to ensure the cluster's competitiveness.

> There is a rising expectation on the role of universities in the field of optical science. The Shizuoka University Venture Partners fund was created to invest in university-based venture business start-ups. The chamber of commerce and industry is disseminating information about successful cases of partnerships and encouraging firms to become more competitive with the purpose of reforming the mind-set of small- and medium-sized firms in the region.
> (JETRO, 2004: p. 46)

By furthering applied research and technological innovation under these programmes, Hamamatsu is working to foster the growth of an opto-science industry as a new industry (JETRO, 2004). 'The Japanese tend to normalise their knowledge among an entire industry. It is not a practice that is illegal in Japan and it certainly is an effective means of intra-industry technology transfer' (Eagar, 1986). Nevertheless, individual firms are cautious about leaking commercially sensitive information to potential competitors.

> We are afraid we cannot release details of our collaborative research with companies from different industries, as contracted. There is a recent example of collaborative development funded by the public funds; this is the Collaboration of Regional Entities for the Advancement of Technological Excellence (CREATE is a regionally concentrated collaborative research project), operated by the JST, where the Research Foundation for Opto-Science and Technology participates as a core organization.
> (Photonics Hamamatsu, personal communication, May 2006)

6.3.3 Vignette: collaboration to create innovative stainless steel textiles

In the creation of the 'stainless series', Reiko Sudo describes the importance of working with artisan practitioners in textile manufacturing companies [small- and medium-sized manufacturers (SMMs)] and writes:

> We worked with micro-fibres developed by a company making radial tyres and spun with stainless steel to create threads. Without the mill's expert technique for weaving at perfect tension without destroying the looms, we wouldn't have been able to create our fabrics.
> (Millar, 2005: p. 18)

Japan has the reputation of supporting collaborative structures and practices. Information sharing in Japan is said to be very efficient and uninhibited because of a high level of trust and commitment to communications (Aoshima, 1993; Berkowitch, 1996; Byrne, 2006; Eagar, 1986; MIT, 2006; NSF, 2006; Rebentisch, 1994; Sako, 2000; Tohoku, 2006; United Nations, 2006). In other industry sectors in Japan, such as engineering, the role of professional associations binds together academics, alumni, and small companies. This creates a mechanism for aligning research funding from government, academia, and the private sector (Eagar, 1986; Fine and Raff, 2001; MIT, 2006; Sako, 2000).

However, as explained in Chapter 3, Section 3.2.2, textile manufacture is a mature and declining sector. Unlike growing industrial sectors, it attracts no special support from government. Despite this, Nuno Corporation initiates many collaborative projects with manufacturing companies. Although there is no special help to sustain any formal professional community in the textiles sector, it is clear that Nuno invests significant time and effort in such collaborations.

HB: Do you think professional communities work together more closely in Japan than in other countries?

RS: I only have experience working in Japan, so it is difficult for me to comment on this question.

HB: Do people in the textile industry belong to a professional association and if so does this enhance innovation?

RS: I do not know what others do, but I personally do not receive any special support, advice, or input from any professional association.

(R. Sudo, personal communication, May 2006)

Collaborations with a diversity of industrial partners enabled Nuno Corporation to create a huge range of innovative textiles. Nuno has achieved a strong international influence and attracts a 'community of practice' made up of inspired students and academics. The approach has given confidence to a new generation; an international movement of creative artists keen to engage in textile innovation in the hybrid space between art and industry. Hence, the popularity of the range of books published by Nuno from 1997 and the high regard for Reiko Sudo's international exhibitions. This is discussed in more detail in Chapter 10 (p. 278).

It is unusual to publish the details of an approach to industrial collaboration with specific descriptions of how an artistic vision was brought to fruition. It may seem risky to provide detailed descriptions of adopting processes developed in different industries and applying them to textile innovation. 'Although along with collaborations always come limitations and imitation, it is very interesting to be imitated. Collaborations help to give us new vision and ideas to deal with imitation' (R. Sudo, personal communication, September 2006). From this perspective, imitation is seen in a positive light and taken to be a compliment and recognition of successful design and a spur to experiment more extensively. This experimental playful approach to innovation is close to the concept adopted so successfully at Google in the USA. It has also been described as 'serious play' by academics at Harvard University (Schrage, 2000). This is described in more detail in Chapter 7 in Vignette 'employee creativity – the object of Google activity' (7.4.2).

6.3.4 Vignette: collaborative information and communication technology (ICT) research project

An HTC project was devised to address skill gaps in local information and communication technology (ICT) companies. These companies were encouraged to share their expertise and collaborate to expand their technological know-how and capacity. A collaborative research project acted as a boundary object to identify skill gaps across the companies and to identify complementary capacity. The research addressed the problem that many jobs for skilled technicians were unfilled. The 'skill gap' was a growing problem and constrained the capacity to innovate, but the solution was not clear.

> Coincidentally the New Technology Institute was commissioning a study by the Sector Skills Council, e-skills UK, to research the supply and demand of ICT skills, so it was agreed to combine resources and to devise a joint action plan on the basis of this research.[2]

To clarify the situation, the HTC2 steering group joined forces with the regional ICT cluster group and the New Technology Institute (NTI) to develop a collaborative approach to commissioning a piece of research.

> The training managers of IBM, Oracle, Marconi (and a representative from the local SME community) have been asked about the key skills shortages. The outcome is that the NTI has developed new curricula to meet those needs via FE and HE courses to remedy those skills shortages.[3]

Three hitherto separate communities or activity systems worked towards a shared object. Combining resources and aims helped to coalesce a shared understanding of the contribution, each of the partner organizations could make to address the region's economic and innovation challenges. It was also the catalyst for more extensive cooperation between the local companies. They decided to work with the HTC on a bigger project to identify local expertise and additional capacity. A network of companies was created that had the confidence and trust to help each other.

Figure 6.3 shows how the research into skill gaps crossed boundaries to bring together sources of funding, expertise, and the users and beneficiaries of the study.

The project formed the basis of more extensive collaborations, which attracted public funds and connected the project beneficiaries to the 'meso-policy enactment level' and to a supportive cluster network.

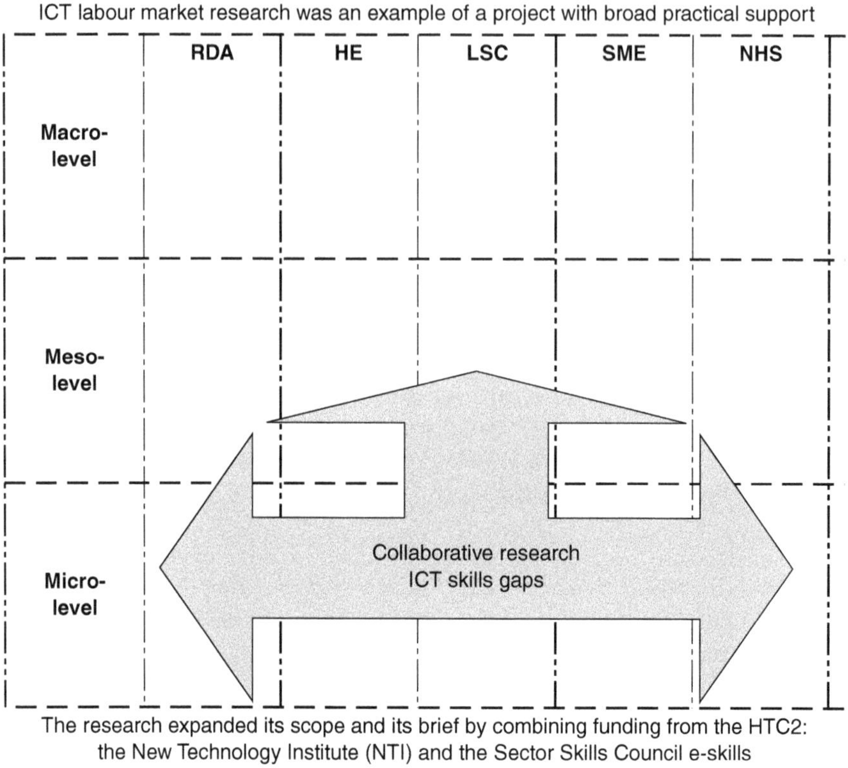

Figure 6.3 Collaborative research into sub-regional skills gaps in ICT.

> It is valuable to call on one another because the Information Technology (IT) industry has lumpy workloads. Swapping business opportunities and looking for synergy, where companies can combine complementary skills can make a real difference to the size of contracts that can be handled. Sharing resources, and specialist skills such as 'C++' expertise is a good thing for the region.[4]

The companies were encouraged to enter new markets based upon their combined skills and capacity.

> We encourage the IT specialists to apply their expertise to other sectors. It's about diversification and the need to know other players join other sectors. We are now at a stage where the project is getting really interesting.
>
> The project created a network of small companies, and this network transformed their capacity to bid for huge public sector contracts, that were out of reach to any of them working in isolation. It began a miniature expansive cycle from which fuller collaboration could grow.[5]

6.3.5 Vignette: learning from a rapid prototyping project in the UK

The next vignette highlights the importance of motivation and attitudes towards collaboration. In some circumstances, collaboration is instrumental rather than expansive and may be more accurately described as a strategic alliance. In such circumstances, individuals and social groups agree to work together for mutual benefit, and so the collaborative practices can be characterized as pragmatic and temporary in nature. This vignette illustrates how something that began as self-interested collaboration transformed itself into something deeper and more sustained.

Initially, a large engineering company wanted to use rapid prototyping equipment available through HTC3 and associated with the Advanced Engineering (AE) cluster project. The aim of the project was to demonstrate how to create prototypes for manufacturing new components in polymers instead of using the traditional pressed metal. The managing director was disappointed when it was discovered that the company did not qualify for assistance. The funding rules governing the HTC project limited them to assisting only small and medium enterprises (SMEs) employing less than 250 people. Although the funding rules for the project made it impossible for the AE team to help the company directly, common sense prevailed and some indirect help was proposed. Conversations with the HTC 'AE project team' led to a mutually advantageous outcome where the very small companies in their supply chain were able to join the AE cluster project. 'The project brought the AE's rapid tooling machine to the attention of small engineering companies who tend to be conservative and traditional in adopting new ways of doing things'.[6] Through the use of rapid prototyping equipment and access to specialized machinery in the university's Advanced Engineering Unit, the AE project was instrumental in changing the culture of SMEs working in the tertiary level of Rover's automotive supply chain. 'In the past most research and development was done at the first tier by the manufacturers, the second and third tiers of SMEs who supply components, are increasingly getting involved in product development work'.[7] The access to specialist tooling equipment and expertise in the universities was beneficial to the small companies who had virtually no budget for research and development. They were suddenly able to develop new products using new equipment and by learning to use new technologies. 'It has changed the way we handle product development, which always used to be done in house as we were at the top level in the "pyramid"'.[8] The technical director of the large engineering company realized that he or she could spend less money on 'in-house' research and development because he or she was now able to pass some of this down the supply chain to the smaller companies. The large company benefited from 'lower piece costs' and 'lower tooling costs' which improved efficiency and profitability because the SMEs in their supply chain gained the capacity to do their own product development. The SMEs in the supply chain also benefited because they had access to the tools to enable them to innovate, and they had the opportunity to work with like-minded firms to develop new products and new markets

These deeper levels of cooperation strengthened relationships between the larger companies and the smaller companies in its supply chain and the university. The project moved the relationship from an instrumental alliance towards a self-sustaining 'partnership'. An *emotional commitment* to the object seems to be crucial to moving forwards from instrumental collaboration towards more sustained expansive learning and cooperation. In terms of ELT, membership of the AE cluster provided access to rapid prototyping equipment and mentoring and was an emotional hook. It acted as a *conscription device* to enrol small companies into collaboration with the university.[9]

This supports the findings of a study of innovation in an engineering company two years ago, which reported that conscription devices, such as emotional ties to new work teams, helped to ensure a process of co-configuration. In the development of new engineering products, 'various design expert' and user perspectives were integrated in the overall design (Ludvigsen *et al.*, 2003: p. 307).

Co-configuration requires trust and the confidence to exchange views, which may not always concur. New insights were created so that the partners had faith in their capacity to achieve more together than would have been possible independently. In other words, there was agreement that they could achieve more by working together towards a new, shared goal. This could be described in terms of ELT as a process by which the boundary object was the tool that started chain reaction of small collaborations leading towards more ambitious expansion and new objects of activity.

However, it is significant that the funding rules caused a problem that confounded the espoused policy aim. Only by working around the rules in an imaginative way, could the HTC project team fulfil the spirit of the HTC policy. This illustrates the points raised in Chapters 2 and 4 about the policy implementation gap.

6.3.6 Vignette: learning from a collaborative visualization software project

The next vignette describes the development of a *boundary object* to connect two activity systems and its subsequent transformation as it was used more widely. Traditionally, if a new product does not meet the needs of users, it becomes a problem and the *object of the activity* for a community of users (as discussed earlier). The problem hinges on the tensions between two different conceptions of knowledge and two distinct objects of activity. The challenge is to connect the two perspectives and find a communication tool; in other words, a *boundary object* that helps the two communities to work together on the problem.

If that object is shared and expanded, the *developers* and *users* can work on new solutions together so that the object is a co-construction and becomes a tool in a new expanded activity system. Co-configuration has the potential to accelerate the development of new products in a single integrated process, because user needs and consumer aspirations are considered at the start, rather than towards the last stages.

> Possibly, it's about helping the customer to specify more clearly what they need and want, when unfortunately they probably don't 100% know themselves until they've got what they **didn't** want, and also it's not always the obvious problem that's in front of them that is blocking their future prosperity [emphasis in original].[10]

An engineering company devised a software tool as a *boundary object* to link the expertise of marketing and design teams with engineering teams. The tool aimed to combine the *user perspective* with engineering expertise to improve the process of new product development. 'OPM is a virtual collaboration tool with visualization tools, such as a three-dimensional computer aided design viewer and the facility to upload drawings'.[11] It turned out that the tool had multiple uses and benefits not only for themselves but when it was adopted by an HTC project (known as the *AE project*); they realized it had enormous potential benefits for all the local engineering companies. The AE project hosted the software on their Web site and made it freely available as a tool for a 'micro-cluster' of local companies.

Although the software was first used as a boundary object, it was transformed by being used in new contexts or circumstances. When it was adopted as a tool in an HTC project, it was used by a wider group of very small companies who lacked their own research and development facilities. It was used as a tool in an activity system of small companies operating in the tertiary level of the Rover supply chain. By making the software available for beta-testing via its Web site, the project enabled small companies to use the software to undertake new product development. The use of the collaborative visualization tool created unforeseen opportunities. It opened up new potential markets for small engineering companies who could act as a 'network' because of the HTC project.

Figure 6.4 depicts this story in terms of ELT and illustrates how a boundary object originated from the need to connect designers and engineers (outcome 1). It was then used as a tool to help very small engineering companies undertake new product development (outcome 2). In beta-testing, the software with these small companies, and the commercial potential of the software, was realized, and an international product opportunity was seized (outcome 3).[12]

ELT suggests that multi-level analysis is important. Taking the *subject perspective* of the company, OPM acted as a boundary object for collaboration between their designers and engineers orientated towards the object of new product development. The adoption of the boundary object as a tool by the HTC project was an opportunity to beta-test OPM software. At the same time, the software was made freely available to small engineering companies to help them to create their own new products.

> It developed as an in-house facility . . . but through working with the Advanced Engineering Cluster Project (AE), we saw the potential to develop it into a commercial product. Without AE we may not have had the courage to develop the software commercially. We now have a proven **'tool'** through

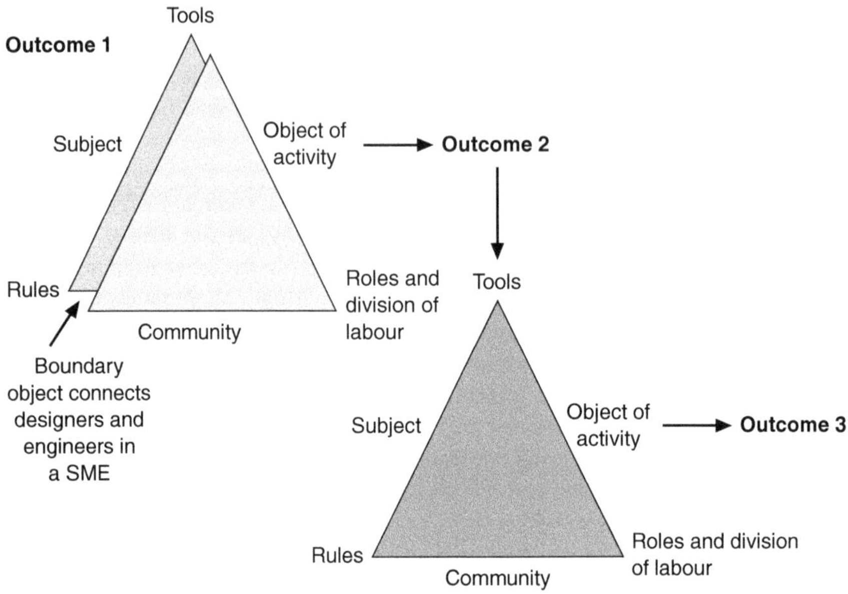

1) The collaborative visualization software (OPM) was used as a boundary object to connect user and developer communities in a single company.
2) It was then beta-tested with a cluster of engineering companies and transformed commercial product.
3) It was later used as tool new product design outcomes in the SMEs at the tertiary tier in the Rover supply chain.

Figure 6.4 Transformation of a boundary object by its use in new contexts.

> operating a free product trial with the small companies in the AE cluster [emphasis original].[13]

Taking the *subject perspective* of the HTC project team, the software acted as a boundary object to connect the perspectives to enhance the process of new product development. The beta-testing provided a free software trial and acted as a *conscription device* to attract small engineering companies to join the cluster. The software helped small companies to carry out new product development despite having no research and development budget of their own.

From the *subject perspective* of the managing director of the company who had created the software, a new object had emerged serendipitously, in that a new commercial product could be launched as a direct outcome of their involvement with the HTC project. Although this HTC project began as an instrumental strategic collaboration, it grew into sustained cooperation.

> We work closely together. We use the AE cluster fairly extensively, the relationship is evolving and we keep talking to one another. AE are developing

> their understanding of the requirements of industry, reaching out to more companies in the area. Sometimes the SMEs don't initially see the potential benefits of the software and its value to a company – We've simply got more structured through working on projects together.[14]

As a result of the HTC sponsoring the beta-testing of OPM software, small engineering companies changed their behaviour and their capacity to carry out research and development. Although the company's initial motivation may have focused on developing their in-house capacity for product design, they were also able to see a new market and seize new commercial opportunities. In other words, working together on shared goals changed individual attitudes, so the perspectives and motivations of the company expanded as new objects of activity emerged.

6.4 Crossing the valley of death

Policy makers frequently describe the process of innovation in three stages. Research emerging from the universities is frequently under utilized and never reaches stage two. This second crucial stage, development is known as the *valley of death.* It is highly vulnerable because it is the domain of neither universities nor industries.

> The path to commercialisation of new technology has three major steps: research, development and innovation. Research is the mechanism by which new knowledge is discovered. Development is the application of this knowledge into technology that solves practical problems. Innovation is the application and commercialisation of developed technology into specific markets, through which industries are born.
>
> (Eagar and Musso, 2003: p. 3)

In the final stage of commercialization, it is clearly the role of industry to invest in developing new products and services to meet the specific need of the available markets, whether these are old or new.

> Entrepreneurs and existing industries have shown a willingness to bear the risk of commercialisation of developed technology, and have built paths and mechanisms, such as venture capital, to encourage such commercialisation. However, there are very few organizations willing to bear the risk of development and even fewer mechanisms designed to encourage it. This is unfortunate, because investment in research is squandered without sufficient development funding to balance the research portfolio.
>
> (Eagar and Musso, 2003: p. 3)

It is this gap that the HTC policy in the UK set out to address. By forging partnerships between universities, private sector industry, hospitals, and public sector research organizations, a variety of financial carrots and sticks were used to motivate 'development'. By contrast, the USA has not experimented with complex government-led initiatives on this scale.

> Development is considered to be the 'Valley of Death'. It has earned this name for two reasons. First, many scientific results go unused because they are unable to attract development funding, and many development projects die early because companies are unable to see the returns necessitated by long development timeframes. Second, academia, where a large portion of Federal research is performed, does not respect or reward development: following a path of development can kill careers. It is virtually impossible to get tenure at a top U.S. research university with development projects.
>
> (Eagar and Musso, 2003: p. 3)

Development projects have traditionally been viewed as the domain of industry, but competitive pressures of the past twenty years have resulted in a business climate that places a premium on immediate profits. While this push improves many aspects of business, it is detrimental to the development of new technology. For various reasons, development periods for certain advanced technologies, such as new materials, can span ten to twenty years (Eagar and Musso, 2003: p. 4). The HTCs used a complex array of EU and UK DTI public funding to bridge this gap. However, the HTC programme and its component development projects were frustrated by over-complicated multi-layered mechanisms. The bureaucracy itself became a barrier to action and was counter productive to changing attitudes and behaviour in the private sector.

By contrast, the Japanese attitude to sharing the costs and fruits of research and development within an industry seems to occupy the middle ground between the polarity of prevailing UK and US attitudes towards public investment in innovation. The critical mass attained in the Japanese model shortens the innovation cycle and brings competitive advantage as products reach their markets more quickly.

The UK case shows that forming a multi-agency partnership, without a clear idea of the value added by each 'partner organization', was detrimental to the effectiveness of innovation. Innovation arises when different organizations act together to create wealth.

> Long-term development projects are a 'valley of death' for many advanced technologies, because there are no clear development channels. Industry cannot afford the risk of 5 to 20 year development projects. Small businesses, which have been the most effective technology developers, lack the resources to even attempt such projects. The culture of academia is skewed heavily toward science, and the type of creativity necessary for development projects is neither encouraged nor rewarded.
>
> (Eagar and Musso, 2003: pp. 6–7)

By trying to break the process into easily understood components, there is a risk that a distorted picture emerges which does not capture the holistic nature of the innovation process. First, the processes of 'technological change and innovation' can be portrayed as a set of distinct stages, each encompassing a set of driving forces and mechanisms. Second, that there is no single instrument or 'common panacea' that seems capable of facilitating the entire process from invention to diffusion. These observations suggest that an innovation-oriented policy strategy should include a carefully selected portfolio of instruments that does justice to all stages in the innovation cycle. The complex inter-relatedness of influential factors calls for a more integrated analysis that allows for 'fuzzy boundaries' (Christiansen, 2001).

Eagar suggests that in order to justify investments in innovation, there have been attempts to devise metrics that quantify inputs and outputs of public and private sector money in terms of return on investment (ROI) in the USA.

- Measuring manufacturing, as a generator of wealth, instead of as a direct employer, to help policy makers understand the true impact of changes in the manufacturing sector.
- Improving continuing education of manufacturing workers, to help improve direct product and process innovation, and to prepare workers for future industries. Every worker should be numerate and literate.
- Balancing Federal research budgets between research and development, so that research expenditures aren't squandered by failure to fully develop the new knowledge.
- Requiring researchers to include potential applications and cost/benefit justification, to ensure a favourable return on taxpayer investment.

(Eagar and Musso, 2003: pp. 6–7)

In the USA, 'innovation' can be characterized as being orientated towards market-led activities of the private sector with strong attention to their ROI. The USA has a stronger track record of commercialization of new technologies than Europe. The commercial success of a technological development results from technology pull in the private sector. In the USA, the technology pull and the clustering of service industries around high-growth companies in Silicon Valley seem strongly aligned so that commercial spin outs of research from academia are brought to market more quickly.

6.4.1 Vignette: the role of venture capitalists in Silicon Valley

The cultural and financial environment in Silicon Valley has proved almost impossible to replicate elsewhere. Despite the widespread adoption of the formula of clustering high-technology companies around sources of research expertise for instance in University Science Parks, the environmental conditions appear to be unique.

> There are numerous informal venues for pitching your company – over breakfast meetings at the restaurants around Palo Alto. The culture there thrives on discussions of business plans over a laptop. The atmosphere in 'The Valley' encourages spontaneous encounters between people with ideas and people with the experience and finance to make things happen.
>
> (D. Aslin, interview transcript, 11 October 2006)

In Europe and Japan governments, agencies have intervened to catalyse tripe helix engagement and to stimulate the development of science parks, 'technopoles' as sites of innovation for companies dependent upon research and new technologies. However, it has been almost impossible to create a climate that attracts large-scale investment of services industries and venture capitalists.

> Silicon Valley has a full range of service industries clustered around the entrepreneurial companies – such as IP lawyers, executive search facilities and lawyers to write the contracts for potential employees – as well as more than 4000 venture capital investors. Nothing exists on this scale in the UK not even in London or Cambridge. Within this compact location you have access to all you need to build great companies.
>
> (D. Aslin, interview transcript, 11 October 2006)

In the UK, banks have been relatively unsupportive of start-up companies. They are highly intolerant of risk and set high interest rates for loans and are quick to withdraw financial support when new companies face problems. This is a very short-term policy, which makes it more difficult for companies to invest in new equipment and research and development. Silicon Valley is special.

> It's a combination of things. Silicon Valley creates a huge amount of innovation and failure. There is access to large amounts of capital investment, and access to the right skills sets, including Universities. It attracts people from Universities around the world. The environment encourages innovation in terms of people willing to work together to create the next new technology.
>
> The secret is not just this and money. The culture is special. The culture tolerates huge amounts of failure. One in ten businesses are successful and one in twenty survives more than five years. Teams move on and start a new company. It is not unusual to find people who have moved on after 6 or 7 failures of new ventures.
>
> I'm not sure if it can be done elsewhere because in Silicon Valley there is no stigma attached to failure – if the failure is of the right kind. It's difficult to copycat a 'culture'.
>
> (T. Foremski, interview, 2006)

The most conventional relationships between large and small companies in Silicon Valley can best be described not as acquisitions rather than collaborations.

Decisions are market driven, and long-term investments in emergent technologies are commonplace.

> Large companies will buy small companies. The Intellectual Property (IP) market is quite healthy. Venture Capitalists will help with mergers to grow a company. Although Initial Public Offering (IPO is equivalent to floating a company on the UK stock market) is difficult these days the regulations are tricky and it's expensive for small companies. IPO can take between 30–40% of the revenue of a small company – so this really amounts to a tax on innovation. Key ways for investors to get money is to sell small companies to larger ones. Even if the small companies fails – they often have good IP assets. Sometimes even another start-up will buy the assets of a filed company and sit on the IP until the time is right to exploit it.
>
> (T. Foremski, interview, 2006)

This approach has potentially negative impacts as well as the more obvious benefits, but this is a huge cultural difference between Europe and the USA.

> There is also a tolerance of failure – and that is different from the UK. I'm always very wary when backing a first time CEO – but I'm particularly suspicious of a CEO without a failure in their track record. Managing in a downturn or adversity is a valuable teacher.
>
> There is a favourable financial environment in Silicon Valley including about 500 venture capital companies with 5000 venture professionals. This creates a huge body of experience to build a company. There is also debt venturing which might be a better option than giving up equity in some situations. It might be a more appropriate to take a loan to buy capital equipment for instance. The banks in Silicon Valley are familiar with start-ups and there is a tolerance and understanding of risk – including clear knowledge and acceptance of the fact that there is a 50% failure rate for start-ups. More importantly there is a willingness by the venture community to intervene early on, if things go wrong. In addition to considering refinancing the start-up, sometimes the solution is to shut down the company, sometimes to reduce the workforce, or sometimes to sell the company on to a bigger company who can take the start up under their wing.
>
> (D. Aslin, interview transcript, 11 October 2006)

Government-led partnerships with the private sector are not seen as sound mechanisms to engineer economic growth. The rules and conditions imposed on government grants are generally understood to inhibit innovation.

> Government has a light touch in Silicon Valley because it's a sustaining environment. Most Venture Capitalists would advise against taking Government grants for subsidising employees or research, unless the Government

> wants to fund research for something that is exactly in line with what the company wants to do anyway. You need to be sure that the company will not be distracted from its primary market and product focus. Even so the grants applications and associated bureaucracy are as horrendous as any in the world.
>
> In the US the scale of ambition and the appetite for business success is huge and lasts even until people are in their eighties. Here you can meet serial entrepreneurs every day that have made lots of money from their start-up successes, but still have the appetite to do another one. All these cultural differences affect the innovation cycle – that's been, quite accurately in my view, described as 'creative destruction'. The boom/bust, success/failure and if it doesn't work, try something else culture thrives here.
>
> (D. Aslin, interview transcript, 11 October 2006)

Experienced operators in Silicon Valley and other innovation 'hotspots' in the world are somewhat sceptical about the capacity of Government to catalyse change. It seems unlikely that the financial climate in the UK could have supported an innovator such as Joe Krauss who has experienced both spectacular failure and spectacular success.

> Well, Joe Krauss the founder of Excite may be a good example. He was in the search engine game way before Google entered the market. He failed to 'progress' against the search engine competition. Excite did not grow successfully and was acquired. Joe then set up JobSpot, which went on to be sold to Google (in November 2006). These were not necessarily failures. The problem was that Excite and JobSpot could not grow a large enough market to compete on their own. So sometimes changing direction or selling is the right thing for an innovator to do.
>
> (T. Foremski, interview, 2006)

Venture capitalists operating in Silicon Valley are highly sceptical about the capacity of the public sector to stimulate innovation. The prevailing view is that the imposition of bureaucracy outweighs any financial benefits from the US Government grants and actually restricts innovation. 'Government's job is to create the conditions where these ingredients can be created, then get out of the way – if it's done it's job, the entrepreneurs and venture capitalists will do the rest' (D. Aslin, interview transcript, 11 October 2006). For instance in the UK, to even begin to implement some of the financial support available in Silicon Valley would require a dramatic change in the attitudes of the UK financial services sector. The vast scale of resources makes the Valley an unbeatable location for new companies to start up and to grow rapidly. 'It really needs a holistic approach. Cultural change is generational and a really long-term endeavour – and it's the culture, infrastructure and business environment that are the ingredients for Silicon Valley's success' (D. Aslin, interview transcript, 11 October 2006).

6.4.2 Vignette: industry scanning and serious play – jellyfish fabric

The case study of the Nuno Corporation shows how scanning innovations in other industries and then applying them in the context of textiles offers significant benefits. It is also an example of technology pull where Nuno acts as a catalyst for change in manufacturing companies. Reiko Sudo always follows the field of industrial design closely, searching for developments that might be applied to cloth. This was how she discovered the use of a fabric in the brewing industry which she transformed into the jellyfish fabric in 1993 (Millar, 2005: p. 29).

> An associate told me about an interesting straining cloth used by beer breweries. The material shrinks nearly 50% with the heat of fermentation, which causes the mash bags to shrink and press out the beer. And when the temperature rises still higher, above 200 degrees Centigrade, the material disintegrates. So we did our own experiments, stitching on heat moulding polyester, and obtained random, almost organic pleats. By going to the factories and seeing these new materials, I could see adapting them to my own use. The manufacturers had detailed purposes for the material, but I just had to transpose it into textiles.
>
> (Millar, 2005: p. 17)

6.4.3 Vignette: metallic plating of textiles

As a student of industrial design, Reiko Sudo wrote 'we were given time to *find ourselves* [emphasis in original] and decide our specialization, which I suppose is rare for a Japanese art school. That gave me familiarity with different materials and disciplines and I kept those diverse connections alive' (Millar, 2005).

> 'Sudo looked to industry for her first collaboration, in particular spatter platting, a method of applying metallic trim to car parts. At that time only one automobile company in Japan was spatter platting and primarily on small surfaces such as doorknobs. With much difficulty Sudo learned to spatter-plate newly emerging micro-fibres, that is 'to bond metal powders to the finest of synthetic cloth'. Her difficulties were compounded by the need to repeat the process three times one for each of three powdered metals-chrome, nickel and iron-which compose alloy stainless steel. Nuno's resources were limited so Sudo raided her own bank account. Believing that textile is a material like stone or glass, she began to spatter micro-powders onto mirror smooth polyester to make stainless steel cloth. The results were astonishing. The cloth was beautiful, billowing, soft and shiny. The finest silver membrane seemed to float on an underpinning of air. It wasn't until 1994 that a Japanese textile factory installed the equipment for metal-plating fabric and the cost dropped dramatically.
>
> (Millar, 2005)

Nuno Corporation takes a unique approach to innovation. It does not conform to a linear model where three stages, research, development, and commercialization, happen in a predictable sequence. Small companies are perceived to perform badly at the development stage, but this is where Nuno Corporation excels.

> Development requires a different type of creativity than science, and that type of creativity is not valued in the current university environment. Small businesses and individuals have proven to be very effective technology developers. Unfortunately, few small businesses can afford to engage in long-term development projects because of capital constraints.
>
> (Eagar and Musso, 2003: p. 4)

Nuno is clearly unusual, because as a small company, they are involved in all three stages of innovation. Experimental projects happen on an ad hoc basis, but there is no overall plan to embed innovation in the Japanese textiles industry. Nuno Corporation makes a sustained effort to pursue creative and artistic challenges, and their beautiful new textiles have an international influence on perceptions of the scope of textile design. The practices at Nuno Corporation show that when artists work with industry, they can make a significant impact on innovation.

6.4.4 Vignette: collaboration on hydrogen fuel cells in California

Until recently, Californians had a long-standing distrust of the renewable energy industry. This has been a major force discouraging and, in some case, even blocking wind, solar, and other ventures from expanding enough to ease the grip of the state's current energy crisis (San Francisco Chronicle, 2006b). According to experts and industry sources, the combined impact of the utilities' stance, the regulatory actions, market forces, and lack of governmental leadership has impeded the use of renewables. All these factors conspired in the past two decades to prevent the renewable industry from producing more than 12 per cent of state energy supplies.

> The struggle California has faced in tapping into clean electricity sources is partially rooted in the state's energy crisis, which still looms over the energy industry here and has slowed the development of all new power. But it also suggests the difficulty politicians, regulators and businesses may encounter as they make the dramatic move away from a carbon-based fuel economy.
>
> (Martin, 2006)

By not developing more renewable sources, the state lost 'a potentially powerful hedge' against its heavy dependence on natural gas and out-of-state producers. Almost one-third of the state's electricity is generated using natural gas.

> 'The renewables have not been promoted heavily by the state or federal policymakers or by the utilities, and consequently the industry has not been

> able to evolve a strong commercial base', said Dan Kammen, an energy professor at the University of California at Berkeley. 'The California situation is enlightening a lot of businesses and individuals about the need for an alternative energy source for backup or primary power', said Jim Kirsch, a vice president and head of a power generation unit at Ballard Power Systems in Vancouver, British Columbia. Many energy experts have long championed hydrogen's potential as a power source – the key ingredient in hydrogen fuel cells that offer a pollution-free alternative to batteries.
>
> (San Francisco Chronicle, 2006b)

Collaboration between the public and private sector with alignment of national and state policies and incentives for business seems to offer the best hope of a solution.

> Oakland Mayor Jerry Brown, . . . said what was needed statewide 'is collaboration of the utilities, government and everyone to create a more sustainable energy resource that does not disrupt the global climate'. It's not enough to just solve this immediate energy crisis: It has to be solved by state and national policies and not by finger-pointing and blame avoidance.
>
> (San Francisco Chronicle, 2006b)

Solutions are already visualized, but as yet the practical steps towards a unified policy to create the desired results have yet to be taken. Robert Stempel, the former chief executive of General Motors, now chairman of Energy Conversion Devices, has formed a joint venture with Texaco to develop solid-state, metal-hydride hydrogen storage systems for powering clean-running vehicles. Ultimately, the idea is to move away from fossil fuels and other traditional energy sources towards the 'hydrogen economy', in which renewable solar and wind generators can be used to produce pure hydrogen fuel out of water. Driven partly by government clean-air standards and the need to reduce hydrocarbon emissions, corporate America has embarked on a crash programme to turn fuel cells into practical products.

> A hydrogen-based commercial backup power system is due out this year from Ballard Power, ranked among the leaders in the nascent fuel-cell industry. The new system is billed as a clean, noiseless alternative to portable diesel generators. Rather than using water to produce hydrogen fuel, however, the system produces its own hydrogen by breaking down an ordinary hydrocarbon fuel, such as propane or natural gas, which the user has to supply.
>
> (San Francisco Chronicle, 2006a)

Pure hydrogen has some ideal characteristics as an energy container, but those same characteristics make it difficult to handle.

> If a practical hydrogen storage system can be perfected, and if fuel cells can ever be mass-produced cheaply enough, today's utility customers would

> have electricity in a stable, portable form capable of being used whenever needed. Imagine city streets full of fuel-cell powered vehicles, neighbourhood-size power plants using hydrogen, and homes and businesses with stacks of fuel cells in the back yard or basement. These could augment and sometimes supplant electricity supplied through the public grid and might even be wired into a computer-guided 'distributed generation' scheme via links to the Internet.
>
> (San Francisco Chronicle, 2006a)

There is at last an emerging consensus that 'hydrogen will be the fuel of the future'. Hopefully, the private and public sector can unite to make that a reality in California.

6.5 Overview

ELT describes expanding the horizontal partnership and infers a voluntary collaboration between partners, each endowed with a similar degree of agency. A *boundary object* can connect perspectives and might take the form of an artefact, a discourse, or a process.[15] Typically, a boundary object might facilitate collaboration between communities of experts from different cultures but with complementary skills (Tuomi-Grohn, 2003: pp. 202–203).

The empirical data revealed a more complicated process of expansive learning where boundary objects act more powerfully when used by a mediator. These projects acted as catalysts for change sometimes leading towards a fuller and sustained expansive cycle. Several vignettes illustrated how particular projects changed practices and focused on instances where boundary objects were transformed when used in new contexts. When adopted by a different activity system and used in a new context, a boundary object can transform into a different element of activity. For instance, vignette 'learning from a collaborative visualization software project' described how software was developed as an in-house communication tool and eventually launched as a commercial product, so it was finally an *outcome* in an expanded activity system.

The data show that boundary mediators have *agency* and can connect the perspectives of activity systems operating at different levels and usually separated by the dynamics of power and politics. They can mediate between perspectives from more than one level, because they understand the *object of activity* from these different positions. Yet, there was no evidence in the empirical data of a mediator being influential if they have no strategic role, in the uppermost level of macro-policy enactment. A mediator can be most effective if they are actively involved as members of the activity systems operating at all three levels of policy implementation. The capacity of a mediator to balance the three contradictory 'objects of activity' seems to depend on their ability to resolve internal tensions between their various identities and motives when acting at different levels in the policy implementation process.

My research suggests that small collaborations at 'project level' can create the foundations of trust for more expansive learning and collaboration. This constructs a history of collaboration on which people can reflect. In turn, successful projects can drive more extensive and sustained collaboration as discussed in the practices described in the UK healthcare sector. My vignettes described discrete projects that act as miniature expansive cycles. These projects acted as catalysts for change sometimes leading towards a fuller and sustained expansive cycle.

Chapter 7 discusses some strong contradictions that emerge in the empirical data. These are presented as stories of dilemmas and how the tensions within and between activity systems are rebalanced. Vignettes illustrate the significant role played by mediators working in different contexts.

7 The impact of complexity and power dynamics on organizational learning

7.1 Complexity of partnerships as a constraint on learning

Over the past fifteen years, there has been a global trend to bring organizations from the private and public sector together to tackle intractable economic and social issues. In Europe, innovation policies were influenced by triple helix theory, and this led to the formation of some extremely complex partnerships.[1] For instance, in the UK, the high-technology corridors (HTCs) involved a vast array of diverse organizations, and lengthy, complicated implementation chains impeded effective communication. This increased the risk of wide gap between the policy intention and what was achievable in practice.

It is possible to view change as a process that begins with an analysis of the 'pre-change situation'. This is often called a 'situational analysis' (Trowler *et al.*, 2002). In the instance of the HTCs, consultants SQW were commissioned to produce a report for the Secretary of State on the probable impact of the Rover Crisis. This was, in effect, an analysis of the circumstances that prompted the policy change. The report analysed chronic features of the situation and long-term concerns.

The analysis of 'conjunctural features' took place later in response to the allocation of new funds from the Department of Trade and Industry (DTI) to address the problem. It outlined how key people and new sources of funding would help the region to resolve the crisis.[2] So, an immediate change was to assist local small and medium enterprises (SMEs) (employing 20,000 people in the Rover supply chain) to diversify from manufacturing automotive components. The HTC policy was a crucial mechanism for transforming the West Midlands. The policy assumed that a knowledge economy could be created through regional public and private partnerships and that local action would be more effective than 'big government' solutions (Connell *et al.*, 1995).

The policy was informed by fragments of ideas rather than by a coherent theory of change. In particular, the empirical data showed that *policy borrowing* from the USA influenced the assumptions about how the HTC policy could improve technology transfer (Halpin, 1994: p. 199). The HTC policy was also influenced by a recent Organisation for Economic Co-operation and Development (OECD) study of regional development and economic growth including European regions

(Goddard and Chatterton, 1999). However, the urgency of the response to the Rover Crisis made systematic theorization and testing of the approach difficult.

Consequently, a highly complex and ambitious change programme was launched very quickly, with little opportunity to test the approach in a small-scale pilot study. Although there was some clarity about goals, there were very limited opportunities to engage all stakeholders in the process of thinking about change (Connell *et al.*, 1995). Therefore, it is not surprising that the empirical data showed that particular stakeholders had different tacit theories and directed their attention to divergent, even conflicting, means and ends.

In the original Rover Task Force (RTF) Report in 1999, the consultants suggested that the success of the HTCs would depend upon a number of critical factors. Nevertheless, only some of the recommendations in the report were implemented. There were discontinuities, inconsistencies, and lacuna in policy that in aggregate caused major difficulties as implementation of the policy proceeded, and the delays described below were typical and had a detrimental effect (SQW, 2001).

> Although the main focus of our work was on initiatives within the corridors, there are also a number of **generic issues**, which should be addressed regionally in order to support growth of new, high technology activities .... A priority is to increase the resources available for commercialisation, including expertise and funding .... More should be done to **increase capacity** both within the institutions themselves and in **firms** that are potential recipients of technology [emphasis added].
>
> (SQW, 2001: p. 12)

The contrast between the espoused policy vision at the top regional level and the pragmatic objectives developed at project level illustrated the problem of the *implementation gap*. However, the extent to which the policy vision was informed by robust theory and was shared by the major partners is an interesting question.

> What intrigued me at the time and what intrigues me in retrospect is that the corridor policy emerged and came to the centre of the economic policy stage in the region **without too much serious debate or evaluation** and the regional economic strategy as it was emerging had **major lacuna**. The corridor notion fitted it well and we saw in really quite a short time, the germ of the idea being translated into one of the central points of regional policy [emphasis added].[3]

Key strategic actors believed that the lacuna in policy could not have been addressed at the time of policy formation, because they were revealed gradually in the developmental history of the HTCs. In the early stages, the precise location of the HTCs was undecided, and their impact was unpredictable. The vision of the RTF out of which the HTC policy emerged was driven by research carried out very quickly by consultants and in which solutions to the region's problems were proposed. The recommendations of the report were adopted with little dissent or time for reflection.

> Once the idea had taken root the RDA in typical style, decided that it would brief consultants to undertake an evaluation to make sure that the idea was robust and then if it was – as it turned out to be – to begin to lay out **an implementation plan** which was essentially – driven by the requirement to make the best of European funds, of the monies we had made available through the Rover Task Force, and the RDA's resources. That drove the second stage of the consultancy report, which was, if you like, the first go at an implementation plan.[4]

In other words, the complexity of the HTC innovation partnership made the process of implementation very challenging. The success of HTC policy was, to some extent, dependent upon the accuracy of the original analysis and the effectiveness of the solution proposed in 1999.

7.1.1 Vignette: practical experience of large-scale innovation

Naturally, there were different ideas about how to set up HTCs, and no single organization assumed a clear leadership role. The HTC partner organizations were tasked with the creation of a 'knowledge economy' that could re-orientate the future ambitions of the region and, at the same time, address the problems faced by traditional manufacturing companies. Although some small, experimental technology transfer initiatives had been successful in the region over a number of years, there was little to prepare the stakeholder organizations for ambitious 'technology push' on such a huge scale. No one in the region had experience of tackling an issue of this dimension or significance.

Successful examples of universities driving regional innovation persuaded policy makers that 'what worked in Massachusetts' will work equally well in an English region. To encourage the establishment of high-growth, high-value companies along the main communication routes around Massachusetts Institute of Technology (MIT) is a very different proposition to connecting rural and urban areas in the West Midlands, in which the economic climate and contexts for innovation were quite different.

> Although there have been good experiences in the States, of technology driven universities, such as **Silicon Valley**, driven by the surrounding defence industries as well as the Universities. In **Boston** Route 128 is a development linked to Massachusetts Institute of Technology (MIT) and it works. But there are very good reasons for these strong 'corridor developments' in the States (and other similar regions elsewhere) [emphasis added].[5]

Originally, the ambition had been to set up technology networks, but as it turned out, three HTCs were proposed in the report, linked to geographical boundaries and in close proximity to research capability.[6] This was influenced by the successful development of a 'knowledge economy' in Cambridge and by the model of a *Technopole* (Oakland, 2003).

The Regional Development Agency (RDA) and their economic development staff favoured the concept of technology corridors, because they had a geographical location and fitted in with their notion that economic regeneration takes place in geographic areas.[7] It was assumed that by bringing resources and the private and public sector together, they would somehow be able to agree common goals and HTCs would be a focus for the development of collaborative partnerships. The HTC policy set out an ambitious programme of economic change. The threatened closure of the Rover manufacturing plant was the 'tip of the iceberg', and it focused the attention of the region on its over dependence on traditional manufacturing and in particular on automotive production.

> I think the region was looking at its two big core problems. One was around regeneration. We had to build something really new that was going to provide high growth for the region (and had) the option of looking at combining property opportunities linked to a knowledge transfer opportunity in some way.[8]

Understandably, there were different conceptions of the problems faced by the region and the solutions to those problems.

> The recently developed task force group had expected the investment to be predominantly associated with Rover – but I think that myself, and others, persuaded the RDA that for every job in Rover there were 3 or 4 jobs in the supply chain. It actually had a significant effect not just upon Rover but upon the automotive supply chain.[9]

Traditionally, private sector investment in research in the West Midlands has been very low, and recent surveys indicate that this will not improve in the near future. The complexity of mixing public and private resources in the RTF funding package complicated the operational decision-making for HTCs. Conflicting rules about eligible activity for the expenditure of European Social Fund (ESF), European Regional Development Fund (ERDF), and DTI and resources from the RDA created difficulties and slowed progress.[10]

HTC policy was based on the assumption that local companies would be willing to engage in the process of change. Success in the long term depended on changing deep-rooted attitudes to risk and private sector investment in research. It depended upon companies wanting to work in partnership with researchers through the 'research and development phase'.[11] The difficulty of making such a significant cultural change was not fully understood. It turned out to be an uphill struggle to give local companies confidence in research as a source of profit, instead of just a cost to the business. Companies were not convinced that 'their investment in research would diminish at the point when their profits rose'.[12]

Under the influence of clustering theory devised at Harvard Business School, which suggested that innovation could be achieved by increasing competitiveness of companies, the DTI had urged the English RDAs to create and support networks or clusters of companies with common interests or needs.[13] Porter generalized from the successful clustering of companies around Silicon Valley, Boston, Stockholm, Northern Italy, and Cambridge and extrapolated this into a general theory of competitiveness.

> A company within a cluster often can source what it needs to implement innovations more quickly. Local suppliers and partners can and do get closely involved in the innovation process, thus ensuring a better match with customers' requirements. Companies within a cluster can experiment at lower cost and can delay large commitments until they are more assured that a given innovation will pan out.
>
> (Porter, 1998)

The RDA pursued the business clustering experiment with both the HTC policy and a parallel programme that developed ten regional 'high-growth' clusters of companies. Substantial resources were committed to these parallel activities over a sustained period of time.[14]

It was significant that no resources were dedicated to facilitating communication and collaboration between the HTC and cluster staff, and consequently, companies received, mixed, and complicated messages about business support. For instance, even the people responsible for implementing these policies found the boundaries between the HTC and cluster policies very confusing.

With the exception of the universities, most of the regional stakeholders were inexperienced in technology transfer. Consequently, many of the key figures involved in the development of the HTC policy were influenced by second-hand accounts of successful initiatives elsewhere.

> A group was established at the RDA who were frankly a little bit **clueless** about what to do with this. Although they had 'so called' policy officers, people with responsibility for policy, I think they really needed the assistance of people who were nearer to the technology than they were. They needed people who could make some suggestions about how the investment could take place. So three groups were established one for each of the corridors, which evolved in the process of the debate, I guess, and the three corridors established their own steering groups – and then the policy framework evolved from there [emphasis added].[15]

The 'Rover Crisis' focused the agencies in the region on concerted action. The corridors were originally chosen partly because they were seen as having some potential to attract and develop more high-tech, 'high value-added' businesses by virtue of the presence of universities, research institutions, or property opportunities. HTC policy became the centrepiece of the West Midlands Regional Innovation Strategy (RIS), and each HTC was linked to the resolution of sub-regional problems and the development of new opportunities. The policy aimed to align the private and public sector to drive technology transfer by bringing the worlds of research and industry closer together. In particular, it aimed to change the basis of interaction between the business community and universities and hospitals in the region.

> It's a very broad agenda. Where technology transfer stops innovation ends. Innovation and technology transfer are often used as interchangeable terms. Essentially it's about taking knowledge from one place (a University or research agency) to the business base. It's about how the knowledge is applied to increase profitability.[16]

7.1.2 Vignette: the impact of multivariate objects

The HTCs were uneven in their potential to act as a catalyst for a sub-regional knowledge economy. They had different research strengths and industrial activity, and resources from the RTF funding package were not allocated equally.

HTC1 attracted the greatest share of the resources because of the presence of two research-intensive universities, a national research centre for defence medicine and a University Hospital.[17] These factors influenced substantial investment in the development of medical technologies and the creation of mediparks and new medical technology and photonics clusters. The Rover plant was located within the boundaries of HTC1. It was also the location of a supply chain of small engineering companies that depended on the fortunes of Rover. The ambitious plan was to re-orientate these companies towards the design of medical instruments.

HTC2 focused on serving the needs of two large information and communication technologies (ICT) and telecommunication companies to develop further an existing sub-regional strength. An objective was to strengthen that and, at the same time, diversify towards the development of a strong digital games cluster.[18] At the meso-level of an HTC steering group, there were powerful factors at work, which shaped the meso-object of activity into solving urgent practical problems and saving the jobs in local communities.

> Major threats to the area … there has been a lot of media speculation about whether the Jaguar factory will still be around. Similarly, Peugeot has 800 employees in the city of Y. They have just set up a state of the art plant in an Eastern Europe country which joins the EU in May next year. Again that's a major threat to the livelihood of the people that work in the Peugeot factory, which will have a big impact on the local supply chain.[19]

HTC3 focused on the practical expertise in working with the SMEs at the tertiary level of the Rover supply chain. The Advanced Engineering cluster in the Black

Country aimed to assist SMEs towards development of new products and markets, which utilized more profitably the skills and capability of the workforce. The polymer cluster in Telford focused on assisting polymer manufacturers to look beyond supplying components to the automotive industry. They sought to utilize the private sector research capability in rubber and polymers towards creating new products and markets.[20] Both these projects were specified in the RTF report (SQW, 2001).

Three issues made the implementation of the HTC policy extremely challenging. First, the scope and ambition was unprecedented, and it was driven by a local crisis rather than by a sustained analysis and theoretically robust rationale for change. Second, the policy was so complex and abstract that it was impossible for local companies to relate to it, so it had to be broken down and translated into meaningful components. Breaking the policy into deliverable parts was important. The whole picture was too difficult for SMEs to grasp, and it was not considered important that some aspects of the policy were not fully understood. It was the role of the upper level to aggregate separate work packages into a coherent whole as illustrated in the following extract:

> Whilst the corridor is concerned with all sorts of technology, we can put this on the back burner and for the purposes of communicating a **simpler story** to the SMEs we can focus simply on one thing at a time. A theme such as transportable treatment is a case in point [emphasis added].[21]

Third, the stakeholders in the HTC multi-agency partnerships had no history of working together. The region had no tradition of collaboration between the public and private sector, and in some instances, some of the key regional organizations did not enjoy cordial relationships. There was no history of multi-agency collaboration in the region prior to the Rover Crisis.

> This is the first time that X City Council and Y City Council have worked together on a major initiative. – Rover was a 'wake-up' call to the region and we are now more aware of the untapped potential of the region.[22]

By implementing the policy via sub-regional steering groups, it was assumed that local needs could be met most appropriately. It was especially important to engage SMEs, and this involved telling them about opportunities to catch their attention. Stories of what could be done with research and development helped to make the HTC relevant to their current needs. 'We have decided to work with a **cake and slice** model [emphasis added]. This gives us the possibility to present one of four main sub sets of high technology'.[23] In the HTCs, it is as if social worlds are arrayed like Russian nesting dolls, one inside the other, but data showed that interactions within and between levels are even more complicated. The communication between activity systems operating at different levels is uneven and almost unidirectional. Communication with key strategic actors at the macro-level in the HTC is most effective where a *boundary mediator* acts as a bridge between the levels.

Their role is most valuable where they connect and expand competing objects of activity at the three implementation levels.

The *micro-level object* of activity was orientated towards developing concrete mechanisms to interact with SMEs. The HTC projects were experiments in *pushing* innovation and technology transfer opportunities towards SMEs (Goddard and Chatterton, 1999: p. 83). Projects were developed to raise awareness and improve the capacity of specific groups of SMEs for innovation.

> Although there was a corridor steering group to formulate strategy and the corridor operations group that brings the ten HTC2 projects together … . Each project has its own steering group and a lead partner organization responsible for delivery and their contract with the RDA or the regional government office.[24]

The object of activity for an HTC project tended to be more concrete than the object for the HTC steering group as a whole. Micro-cluster projects were devised in one HTC to target assistance to SMEs. 'We have taken the "High" technology focus down a notch. We provide the expertise, not of the original developers, but of people dedicated and outward looking who are able to meet the needs of the local SME'.[25] The urgency of the need to take the funding opportunity available from the RTF funding hastened the development of policies that otherwise may have taken many years to develop. The discontinuity of practices operating at different levels limited the potential *vertical expansion* of the object. Three competing objects evolved but the consequences were unforeseen and there was no consideration given to integrating or expanding these objects through mediation.

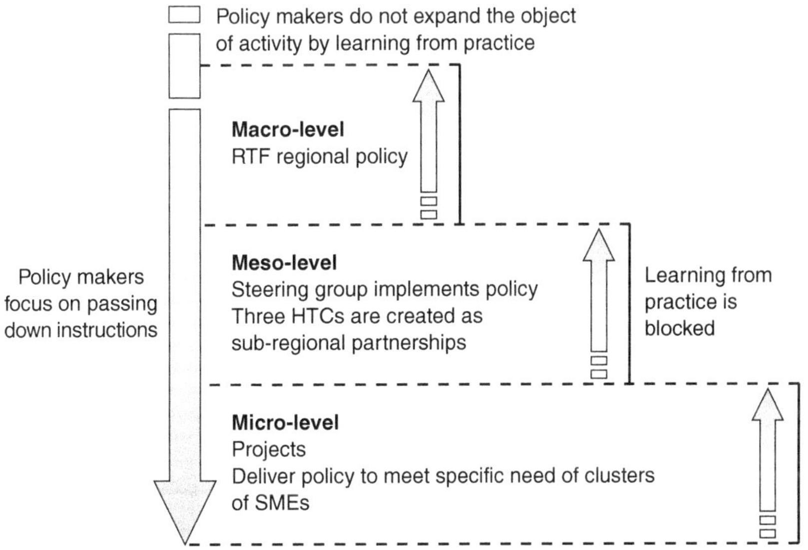

Figure 7.1 Constraints on expansive learning between levels.

Figure 7.1 depicts the situation in the UK Rover case study and illustrates how boundaries between levels resist communication from micro- and meso-levels. Not only does this restrict expansive learning, but it highlights the lost opportunities for the policy trajectory to be informed by feedback from the people who work most closely with SMEs. Given the importance of SMEs to the modernization of the regional economy, the figure points to some serious problems in adapting policy in the light of practical understanding of its effect on those it seeks to help. The structural characteristics of HTC policy implementation blocked feedback from the practice at the micro-level.

7.2 The importance of flexibility and autonomy

This depiction of boundary crossing and expansive learning in Figure 7.1 explains some things about how organizations interact – but it also begs some questions. For instance, 'What happens when there are many different communities, organizations or activity systems and some are more powerful than others?'

By contrast to examples of innovation catalyzed by public funding in the UK and Europe, I will now turn to instances of small-scale collaborations initiated in Japan. Individual companies can operate with a degree of speed and autonomy that gives an enviable degree of flexibility. Vignettes from US and Japanese case studies show the effectiveness of innovation when a company has the autonomy and resources to set up a tightly focused collaborative project with clear objectives.

7.2.1 Vignette: balancing power dynamics in an architecture project

This vignette sketches the characteristic role of a powerful mediator who is able to work across different activity system or communities of practice (COPs). The effectiveness of a mediator to facilitate innovation depends on their capacity to understand how to operate in a different world. Mediators need to have a track record in another industrial sector because they are required to translate practices and applications of technology from one context to another. This example shows how an autonomous small company can innovate by applying technologies from one industrial sector to a new context.

> Nuno has invented over 1,500 fabrics and has been instrumental in changing attitudes so that textiles are considered at architectural development stages of new buildings.
>
> Toyo Ito invited Nuno to collaborate on the Matsumoto Performing Arts Centre. Ito had clad the building with GRC fibreglass reinforced pre-caste concrete inlaid with a special design glass. Sudo and her 'interior design co-ordinator' developed a fabric that suggested a smaller version of the glass inlay pattern of the GRC. 'The curtains create an additional skin to the building that references its outer covering and integrates hard and soft elements of the interior'.
>
> (Millar, 2005)

The Director and Head Designer of Nuno Corporation, Reiko Sudo had the experience and skill to act as a mediator in a major collaboration with architects. However, the relative size of collaborating companies can affect the dynamics of power. Consequently, Nuno finds it more difficult to work with large powerful companies that have a preconceived idea about what they want from collaboration.

> Yes, we try to make some new forms of textiles … . But when we work with such large companies it's like they have a kind of limit. If we see that they have another kind of possibility (they are reluctant to explore a different idea) … . If it is a smaller company it is easier to collaborate. With a larger company, they may not be so free and open to possibilities.
>
> (K. Otani, personal communication, 30 May 2006)

Powerful companies such as automotive manufacturers tend to be unwilling to go with the flow of experimentation. 'In Japan, design, manufacturing, marketing and business tends to be kept completely separate, although the people in these disciplines work closely together and there are strong links between them' (R. Sudo, personal communication, September 2006). However, this example contradicts the view of Berkowitch (1996) who suggests that COPs are much stronger and connections between them are more effective in Japanese industry. The conclusions he formed were based on his studies of Japanese engineering companies. The opportunities and motivations for collaboration in the creative industries are not the same. There are some important differences in the way designers describe the process of innovation.

> We at Nuno never think about 'technology transfer'. We have more than forty different manufacturing subcontractors all over Japan, and we have a very tight partnership with each one of them. Everything we do reflects these partnerships with helpful manufactures that have a deep understanding of creative thinking. Not all of our subcontractors necessarily employ high technology, but technology is always involved during the design stage of our fabrics.
>
> (R. Sudo, personal communication, September 2006)

Nuno has a very interesting approach to innovation, which focuses on taking risks and discovering new artistic possibilities, by collaborating with carefully selected partners that have complementary skills.

> We are a design company, and all of the manufacturers that help to produce our textiles contribute to our creativity. It is really a collaborative effort between Nuno and technician or artisan in the factory who understands our design concept.
>
> (R. Sudo, personal communication, September 2006)

7.2.2 Vignette: how a mediator empowered a weak activity system

My case study of collaboration between public and private sector organizations in California highlights some interesting issues. Reaching agreement on the issue of 'renewable energy' was not easy, and a skilled mediator was needed to reconcile differences between perspectives (or activity systems). The following extract shows how the mediator facilitated debate, to reach a consensus. However, at the same time, he or she was able to document and formally recognize a difference of opinion, held by a minority group.

> Do you have a clear procedure on agreement between different kinds of partners?
>
> For large collaborative groups that study controversial issues … we have a very clear idea of what we mean by consensus. We define it as unanimous or general agreement. General agreement means that decisions do not need to be unanimous, but it does mean the prevailing opinion, which is much more than a majority. Thus, we do not allow a small minority to dictate the results of collaboration. However, the collaborative group pays close attention to minority views. We allow the formation of a written minority viewpoint in the final report. This process takes a long time, but it usually does result in a strong but equitable group agreement on an issue and a clear view of how to proceed.
>
> (M. DeAngelis, interview transcript, 12 October 2006)

In other words, a skilled mediator can enable the voices from a relatively weaker activity system to be heard. This is a long-term approach to partnership that respects all the participating individuals and organizations and, yet at the same time, moves everyone forward by a decision-making process that everyone believes to be fair.

7.2.3 Vignette: the unique nurturing environment in Silicon Valley

In the USA, the practice of clustering has been market led and mutual support has emerged in all sorts of interesting ways that do not necessarily confirm to Porter's (1998) theory.

> What has happened more is that service companies grow around others with needs. In general start-ups tend not to collaborate with each other. They are resource constrained and have very specific objectives – so dealing with another similarly constrained company can be difficult. If they have similar needs, such as a manufacturing facility that they might all share, then that make more sense. Most partnerships tend to be set up between small and large companies – most start-ups need a big brother or big sister to help them succeed.
>
> (D. Aslin, interview transcript, 11 October 2006)

The UK and Japanese governments have tasked departments to stimulate regional clustering of companies with similar technology needs. However, government sponsorship of clusters risks embedding them in a bureaucracy that can constrain their flexibility and spontaneity and inhibit the very innovation it seeks to encourage. 'To supplement the decline in the number of firms, establishment of university-launched venture businesses as new actors in cluster development is just beginning' (JETRO, 2004: p. 43). This could be described as an amalgam of the triple helix concept of partnership and the concept of clustering, which maintains that bringing together companies with similar needs will result in the collective having greater competitive advantage. This suggests a particular notion of partnership being based on mutual weakness, rather than logical synergy between companies. It does not necessarily coincide with the characteristics of a good partnership.

> Well there is a higher frequency of partnerships between start up and big corporations – usually focused on getting the technology developed and to market. Some big companies are well known for crushing start ups – not intentionally – but say just by sending in their top ten scientists for a week to find out what the start-up does – it distracts the company too much at a crucial stage. The more astute big corporations look at resources of people and style of engagement, so that the life isn't crushed out of the relationship.
>
> (D. Aslin, interview transcript, 11 October 2006)

The holy grail of the 'triple helix' policy is the closer integration of universities, industry, and government, and this has certainly been possible in the California. However, in the USA, the process is market driven and independent of public funding and the complex bureaucracies associated with public sector partnerships.

> The universities particularly the Universities of California and Stanford are very orientated towards the start up world .... There is a much tighter integration between the universities and the Venture Capitalist world so that commercialisation can happen where it makes sense.
>
> (D. Aslin, interview transcript, 11 October 2006)

There are also significant differences in the attitudes and expectation of people working in Silicon Valley.

> Well I think people are more willing to line up behind a common goal. Although translation of the terms or vocabulary used by a bio-scientist and say an engineer may still be a problem in getting effective communication in a multi-disciplinary team, it can usually be overcome.
>
> But the real difference is in the attitudes of people. In the UK when discussing a new position or project with an engineer, they would say something like 'why should I make the intellectual engagement with this – I need to see how the technology will challenge my skills'.

> In Silicon Valley the conversation with a similarly qualified engineer would focus on entirely different things. In this region an engineer is likely to say something like 'I can see how I could make a contribution to the company .... Why will this product be successful, how will we make money from this and what would be my share?'. In my view the limited financial literacy in the UK is a real barrier to success.
>
> (D. Aslin, interview transcript, 11 October 2006)

Innovation is truly a complex multivariate problem. There is a big risk in taking part of the recipe and expecting it to work. It is very important to take a holistic view of the contexts in which innovation flourishes.

7.2.4 Vignette: artistic goals and serious play

Reiko Sudo approaches what others might call 'innovation' with an attitude of 'playful experimentation' and tolerance of risk. Like many creative artistes, she values lateral thinking and the freedom to try things that may make no commercial sense at the time. Sudo has sustained her identity as a practicing artist and although she leads a successful international company she is quick to point out that she is not interested in technology transfer but in creativity. 'We, at Nuno, are involved in many different collaborative projects with architects, as well as interior and fashion designers. NUNO and I are more concerned about "creative and aesthetic aspects of our work", rather than the technology' (R. Sudo, personal communication, September 2006). This philosophy has some resonance with the concept of 'serious play' popularized by academics at Harvard University (Schrage, 2000). This approach accepts the risk of failure and that there is no guarantee of commercial success. Playful experimentation is an unusual and effective solution to the problem of 'motivation for innovation'. Nevertheless, Nuno has a long-term vision of innovation and invests substantial time and resource in working with partners in 'the development stage' often described as 'the valley of death'.

7.3 Mediators as mini-maestros of networks

Mediators can act as co-ordinators of collaboration between different types of partners. In this sense, they are like the conductor of an orchestra, in that they bring together the various contributions of different instruments. Each player retains their own instrument with its characteristic sound, but the conductor directs the players to work together to create a sound that exists in his or her imagination.

> As the process of disintegration and reintegration continues, it is becoming clear that the emerging aggregate players will become companies of the future. These mini-maestros bring innovation and efficiency to the network by orchestrating the flow of goods, information and funds between multiple entities and by dynamically reconfiguration the network. Much of the

> competition in the business world will centre on gaining and maintaining the orchestration role for a value chain or industry.
>
> (Bitran *et al.*, 2006)

It is evident that Nuno facilitates a rich diversity of collaborations because Reiko Sudo has an excellent understanding of how to develop methods of working with industrial partners from different sectors. In this sense, Nuno has the capacity to be a conductor of networks.

> Clearly, becoming part of the network is essential; yet becoming the conductor of the network will be even more critical. Looking forward, it is important to understand the impact these aggregators have on the local economies where they operate. Such an insight is instrumental in comprehending the development of trust, quality of service and sustainability in networks made up of small businesses in developing markets.
>
> (Bitran *et al.*, 2006)

7.3.1 Vignette: the capacity of designers to act as mini-maestros – holograms

Flashes of insight characterize the inventiveness of Sudo's textiles. Often, these insights are sparked by unusual connections with processes used in other industrial sectors. While accompanying a friend on a visit to a research centre specializing in holograms, some new possibilities for textiles came to mind. Afterwards, when the friend suggested it might be interesting to make moving holograms, Nuno took this as a serious research challenge. 'This stimulated ideas about flight and led the development of Nuno's highly successful feather fabric series of textiles .... Well, the researchers I met just laughed at this. But it started me thinking about fabric that had motion in it' (Millar, 2005: p. 20). Despite this discouraging response to the initial idea, Nuno's research progressed into the commercial development of a series of textile designs in beautiful fabric that conveys a sense of movement in space and light.

> Feather Flurries was created in 1993.... As the sheer moiré-patterned silk organdie is woven, ... and the loom is automatically stopped and individual duck, guinea or peacock feathers are placed by hand into pockets in the double cloth. The white version reminds one of a wisp of cloud that has captured white duck feathers in its folds. In black with iridescent blue peacock feathers, the textile suggests the glamour of another more elegant era.
>
> (Millar, 2005: p. 29)

While more than a dozen members of Nuno staff create textile designs, Sudo takes responsibility to decide which designs are moved into production. In addition, every fabric has a name, often based on the story behind the fabric. Sometimes, it refers a technique or an idea or a theme (Millar, 2005: p. 17).

7.3.2 Vignette: collaboration in the creation of eco-textiles

During the past twenty years, the scarcity of natural resources in Japan coincided with a second problem. This was the growing problem of the disposal of synthetic materials. Many industrial processes and modern packaging created plastic waste, and disposal was becoming increasingly difficult. Attention was focused on the need for sustainable manufacture and ecological design. This was the motivation for an important international collaboration 'to create eco-friendly textiles'. In theoretical terms, this could be described as the 'object of activity'. As part of Nuno's routine scanning of innovative processes and products in other industrial sectors, an opportunity to respond to these environmental concerns occurred. Designers collaborated with US scientists and technologists over a sustained period of time and numerous projects to create ecologically sustainable textiles.

> Ever conscious of the fragility of the natural environment, she began to search for ways to create recycled fibres to replace non-renewable oil-based fibres. The American agricultural conglomerate Cargill had developed a new natural polymer: made from corn starch and other grain by-products called Modified Starch Plastic. In 1999 a Japanese concern rendered the fibre into a useable thread. According to Sudo it was 'useable but still not very practical'. Although biodegradable, the threads and fabric were impossible to dye, limiting its use to air filters, landscaping earthwork tarpaulin, fishing line and the like. In 2001 Kiryu Textile Laboratory invited Nuno to collaborate on finding a way to dye this new plastic. This was successful and led to a high profile exhibition and the development of other biodegradable textiles made from materials such as charcoal and brush bamboo.
>
> (Millar, 2005: p. 32)

7.3.3 Vignette: combined impact of mediators and boundary objects – NOVUS

Some of the innovation projects devised in HTCs employed skilled individuals employed by universities to act as 'technology translators'. They were tasked to build conceptual bridges between the company with a technological need and a researcher with a new product or process. The translation role was integrated into the NOVUS project from the start.

These 'technology translators' were also effective as boundary mediators because they were well connected to strategic networks. Their role was crucial in creating a bridge between the micro- and meso-levels. Although the boundary mediators did not sit on the steering group, they communicated issues and problems upwards to the macro-level via powerful colleagues in their university. The *agency* of a boundary mediator is not necessarily a black and white issue, and a 'network of strategic support' can enhance their effectiveness.

> NOVUS acts as a **translator** between the expert inventor and the beginner or new user, in terms of the technology.[26] Sometimes that involves asking naïve questions and also the 'what if?' question. Getting both parties onto the same wavelength is important [emphasis added].[27]

Figure 7.2 shows how the NOVUS project used a combination of boundary mediators and boundary objects to bridge the conceptual gap between communities of experts in SME and university activity systems. The boundary mediators had access to a suite of NOVUS project tools, which included financial resources, technological demonstrations, and university research facilities. The mediators used these tools as 'boundary objects' to bring the perspective of SMEs and the universities closer together.

The boundary mediators were alert to the risks of presenting a company with a technical solution that was too radical because this would be outside their *collective zone of proximal development* to use the terminology of activity theorists

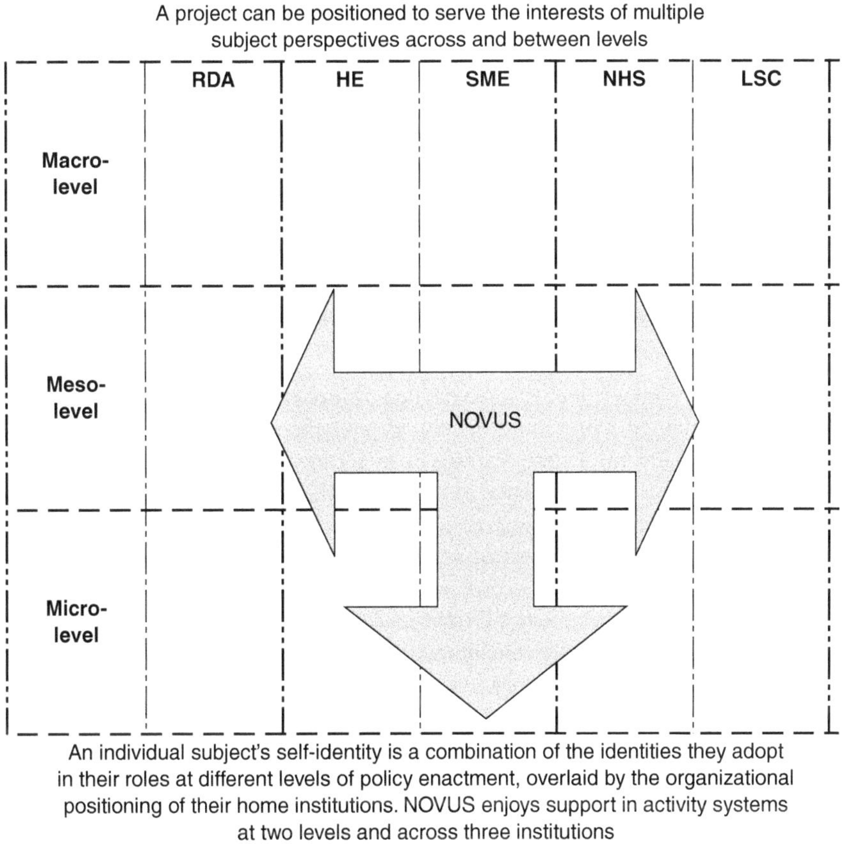

Figure 7.2 How the NOVUS project links users and developers.

(as discussed in Chapter 2). If the change proposed was too far from their current practices, it failed to engage them, and if it was too close, it failed to challenge them to innovate. In developing effective mechanisms for helping SMEs, it was often successful to propose modest changes initially and until the personal relationships were well founded and robust.

> It's about keeping your mind open and not being blinded by the novelty of the invention. There is a risk of a solution being too advanced and not what the company needs. I think users and developers have better conversations with facilitators who talk common sense.[28]

For a small company, the HTC policy would probably be perceived as a *tool*, as one of several programmes that might help their business. By contrast, for the regional policy makers, setting up the HTC steering groups was the object of their activity. 'I think they would view technology corridors in the same way they would view the majority of regeneration devices as being largely irrelevant to them up to the point at which they are suddenly in crisis'.[29] In essence, boundary objects are transitory tools that can be adapted to new situations. Boundary objects can work at different levels of activity and can be reshaped and moulded by new practices as a result of crossing boundaries (Ludvigsen *et al.*, 2003: p. 308). So, as HTC projects crossed boundaries, they changed the relationship within and between activity systems. It is possible to think of 'the NOVUS concept' as acting as the object of activity in one activity system, but at the same time, it acts as a tool in a related activity system. Indeed, the transformation of one type of element of activity into another can seem bewildering.

7.4 Conscription devices and emotional commitment to the object

The emotional commitment to an expanded object seems to be the key to enacting complex change. For a multi-agency partnership such as the HTC, the emotional commitment to change anchors it to future practices and influences its long-term viability. It also relates to the emphasis in expansive learning theory (ELT) on the capacity of *conscription devices* to enrol people with an *emotional commitment* to the object of activity. 'Change is not necessarily sustainable and a project may not be "anchored" in the organization … . Anchoring depends upon the investment of time and energy and a conviction that the project has developed improvements worth maintaining' (Tuomi-Grohn, 2003: p. 228). Medical and healthcare issues seem uniquely capable of aligning motivation and activity systems towards a common object. Expansive learning theorists, particularly in Finland, have researched the healthcare sector intensively. It is very easy to understand why different experts would be able to work together towards improving patient health and 'well-being'. It is probably more emotionally engaging than motives associated with profit or technological improvement.

7.1.1 Vignette: uniting communities in collaborative public interest research

The Sacramento Municipal Utilities District (SMUD) commissioned a small regional project in the State of California. It united environmental lobbyists, householders, local government, and energy utilities in pooling their resources to sustain public interest research The project was described as a *collaborative* and achieved the desired outcomes, by investing a significant amount of time and effort in building relationships between the various partner organizations. The project was unusually successful in aligning the voluntary sector, with public and private sector interests. It was successful despite the overwhelming profit motive of the shareholders of privately owned utilities. 'Private interests in a competitive market do not always reflect the pubic interest. While I'm not an economist, economists do talk about failures in the market based on many factors including externalities and public goods' (M. DeAngelis, interview transcript, 12 October 2006). Plans by private sector utilities parties for a pulverized coal-fired power plant, which could produce cheap electricity leading to profits, had to be balanced with environmental concerns. The public sector and environmental lobbyists were concerned about the impact of the project on global climate change. From their perspective, the large volumes of CO_2 emitted would have a future 'environmental cost' that was not reflected in the current electricity price.

> This example is a significant public issue and is very real for California today, There is plenty of low cost coal in the Western U.S. and with high natural gas prices today, and there are many proposals to construct new pulverized coal plants to supply electricity for California. The California State Government has reacted to this issue by the California Energy Commission (CEC) setting a carbon energy policy standard at the emissions of a state-of-the-art gas-fired combined cycle plant for all new baseload energy supplies for California. The State Legislature passed a bill (SB 1368) to endorse this CEC policy, and the Governor signed this bill into law in September of 2006.
> (M. DeAngelis, interview transcript, 12 October 2006)

Views in California began to change dramatically since the Enron crisis created an artificially high price for electricity and caused power cuts throughout the state. This coincided with a growing appreciation of the problem of global climate change and began to move people away from 'entrenched positions'.

> In the early 1990s investor owned utilities like the Pacific Gas & Electric Company were pressed to do R&D. When electricity restructuring happened they chose to reduce their public interest research. So I formed a '**collaborative**' with environmentalist groups and we went through a two-year process to understand the investment of $240m in public interest research.
>
> We found that declining (public interest research) in investor owned utilities was being forced by preparation for the competitive market. There was

> lots of opposition to the collaborative. But we were able to focus on the facts and issues – so in the end we had a collaborative research programme based on the wish to preserve public interest research [emphasis in original].
>
> (M. DeAngelis, interview transcript, 12 October 2006)

Both the public and the private sector are now interested in renewable energy sources. However, it is possible that the consensus reflects an array of motivations. In this sense, the 'expanded object' embraces those people who want to safeguard the planet and others motivated by a desire to use domestic solar and wind power to reduce their dependence on large energy corporations.

7.4.2 Vignette: employee creativity – the object of Google activity

Google began as a research project in January 1996 by Larry Page and Sergey Brin, two PhD students at Stanford University, California. They hypothesized that a search engine that analysed the relationships between Web sites would produce better results than existing techniques (existing search engines at the time essentially ranked results according to how many times the search term appeared on a page). The company came to prominence at the height of the 'dot-com' revolution, which can be described as a speculative bubble from 1997 to 2001, during which stock markets in Western nations saw their value increase rapidly from growth in the new Internet sector and related fields (Wikepedia, 2006a).

In March 1999, the company moved into offices at 165 University Avenue in Palo Alto, home to several other noted Silicon Valley technology start-ups. After quickly outgrowing two other sites, the company settled into their current home in a complex of buildings in Mountain View at 1600 Amphitheatre Parkway, in 2003. The complex has since become known as the Googleplex (a play on the word googolplex, a '1' followed by a googol of zeros). Silicon Graphics leased the buildings to Google.

> I think they got lucky at a key stage in their growth. Most successful companies have luck. Google figured out the most common processes on the net for finding stuff. Searching had been commoditized so much there was no money in it – Google came up with something that was quicker and cheaper than anyone else . . . They took the chance to create value and built a business in a more profitable way than other search engine companies. Google figured out how to sell advertising via auctioning in a market where it could use a self automated system. Google use machines wherever it can for instance news clippings are harvested by Google computers not people. The company specialises in building the most efficient cheapest computer system in the world. It can use its cheap infrastructure for running various services. It changes all the time, Google licences technologies, and acquires technology. It bought advertising technology from Yahoo but has used it much more effectively and made more of it than Yahoo.
>
> (T. Foremski, interview, 2006)

Currently, the company has approximately 8,000 employees and is based in Mountain View, California. Google's services are run on several server farms, each consisting of thousands of low-cost commodity computers running stripped-down versions of Linux. While the company does not provide detailed information about its hardware, a 2006 estimate consisted of over 450,000 servers, racked up in clusters located in data centres around the world (Wikepedia, 2006b).

The company policy encourages employees to retain the culture of creativity and experimentation that is characteristic of the university sector but rarely found commercial companies (Schrage, 2000). This apparently altruistic philosophy has resulted in huge commercial success. Many of Google's most innovative ideas have arisen from the 20 per cent time allocated to 'serious play'.

> Every Google engineer is encouraged to spend 20% of their work time on projects that interest them. Some of Google's newer services, such as G mail, Google News and Orkut, originated from these independent endeavours. In a talk at Stanford University, Marissa Mayer, Google's Vice President of Search Products and User Experience, stated that her analysis showed that half of new product launches originated from 20% time.
>
> (Wikepedia, 2006b)

Marissa Mayer was previously the director of consumer Web products at Google and has spearheaded almost every user-interface change to the Google Web site 2000. She also teaches computer programming at Stanford and in recent interviews commented:

> It's also important to design for users. We look at the user perspective, not at the how can we make money perspective. If there is a commercial element, we make sure that it's really visible. We try really hard to look at products in such a way so they're how you want them to be laid out.
>
> (FastCompany, 2004)

People have commented that Google have made a very smart move in encouraging employees to spend 20 per cent of their time on their own projects (FastCompany, 2004; Schmidt and Varian, 2005; Sullivan, 2004).

> Communication is key. When you have so any engineers working on so many projects, you need to have tools that help you self-organize ... . We also use them to communicate project status. In addition to the core group of people who organize these pages, people working on other projects can also easily find these pages.
>
> (FastCompany, 2004)

The tools originally developed to help Google employees innovate are then made available to the public who use them in different ways and send feedback to Google, thus spurring them towards further innovations. In this sense, tool is transformed into a new service by being used in different activity systems. Keeping teams and partnerships simple and focused seems to be a recipe for successful innovation at Google.

> Google News took a team of 3–5 people. If you take a pool of engineers and put them in teams of three, you can actually do 100 projects. Small teams can self-organize quickly, they can finish projects quickly, and they can move onto new projects quickly.
>
> (FastCompany, 2004)

The clarity of this approach points to some reasons why complex multi-agency partnerships may not be the best vehicles for innovation.

> Yes, it's smart because Goggle can outdo everyone else. It has the largest most scaleable company platform. Asking their employees to move forward – inventing their best ideas is a good move. The way they do this is in self-organizing teams.
>
> The process creates innovation.[30] Innovation develops naturally. Google has the best PhDs in the world. It's a melting pot for teams to form and collaborate. The problem is that producing lots of different products creates the challenge of how to monetize them.
>
> (T. Foremski, interview, 2006)

However, this business model is not welcomed unanimously; some observers watch the rapid dominance of Google with some trepidation.

> Google has a strong business and millions of spin outs – so it does not need to 'monetize' all the ideas that are generated. The question may be 'Is it fair to sit on things?' It's a bit like Microsoft putting competitors out of business by the dominance of its operating system.
>
> (T. Foremski, interview, 2006)

Venture capitalists, and other experienced players in Silicon Valley, are convinced that this is a highly effective hook, which might be described in terms of activity theory as a conscription device that strengthens employee commitment to the company. People value it highly, and it is a very smart way of operating (D. Aslin, interview transcript, 11 October 2006). Google has also collaborated successfully with large and powerful corporations in the media and entertainment industry. For instance, in 2005, Google announced a long-term research partnership with NASA which would involve Google building a one million square foot research and development unit at NASA's Ames Research Centre. NASA and Google are planning to work together on a variety of areas, including large-scale data management, massively distributed computing, bio-info-nano convergence, and encouragement of the entrepreneurial space industry.

Google unveiled an expanded partnership in 2005 with Time Warner's AOL unit, which included an enhanced global advertising partnership and a $1 billion investment by Google for a 5 per cent stake in AOL. Google also recently formed a partnership with Sun Microsystems to help share and distribute each other's technologies. In August 2006, Google signed a $900 million deal with News Corp's Fox Interactive Media unit to provide search and advertising on MySpace and other News Corp Web sites including IGN, AmericanIdol.com, Fox.com, and Rotten Tomatoes (Wikepedia, 2006b). In October 2006, the purchase of YouTube added another asset to the Google portfolio, which has huge potential.

> YouTube was not acquired for its technology. Google got a new community. They gauge how much money could be gained per customer. They 'monetized' the YouTube community more efficiently than anyone else could. It gives them huge potential revenue provided that it can continue its popularity.
>
> (T. Foremski, interview, 2006)

Even though Google is a huge corporation, there is much evidence of in-house innovation by small project teams. Smart internal collaborations and formal acquisitions continue to diversify Google's business base, but there is a corresponding risk that exponential growth may change the corporate culture (or lack of it) that characterizes Google.

7.4.3 Vignette: collaboration between physicians and automotive designers

In the West Midlands, the small engineering companies in the Rover supply chains were encouraged to diversify into new markets such as the global market for medical instruments and the UK public health sector where collaboration would be beneficial to both sectors in developing new solutions to rising costs of specialized equipment for physicians and devices for patient therapy.

This required cultural change in the private and public sector because success depended on finding new ways of working together to utilize new technologies and product development strategies. In particular, this challenged traditional procurement practices in the healthcare sector. Healthcare managers had never before considered commissioning medical product design from local SMEs. It also involved shifting the 'mindset' of local SMEs and altering the attitudes of healthcare professionals to align public sector agencies to support them in the process of change.

This project acted as a boundary object to link a community of automotive designers with physicians who wanted to collaborate in enhancing diagnostic exercise machinery for patients suffering from physical disabilities.

It was an *expansive realization* that another sector could assist with product development because it was better equipped and had unimagined expertise. Through involvement with the HTC steering group, a key person in a major NHS Hospital saw untapped potential.

> To me it's about taking technology from one firm to another – improving awareness of the latest technologies. It can also be about transferring technology between sectors. An example of this is medical technologies that could be far better developed in the car industry because of their capacity in new product development.[31]

As a direct result of the Rover Crisis, automotive designers were invited to work with a medical team to see whether they could adapt some equipment. An HTC project was a mechanism for collaboration to re-design a machine to the special needs of patients suffering physical trauma. Figure 7.3 shows how HTC policy encouraged new kinds of organizational partnerships including alliances with a new RDA, the healthcare sector, the defence industry, and city and rural councils. The project connected two different cultures and mediated between individuals and groups from different 'COPs' or activity systems. It expanded the object of activity as shown in the figure.

This was an example of co-configuration by experts with complementary expertise. Doctors and physiotherapists collaborated with industrial designers from the automotive industry who used their skills and equipment to solve a new problem.

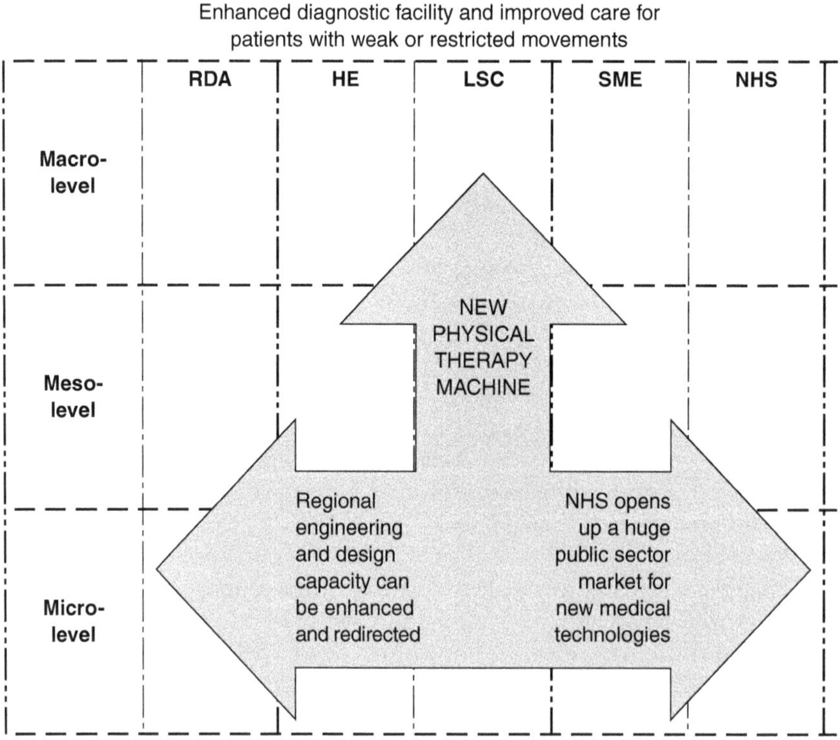

Figure 7.3 Collaboration between physicians and automotive designers.

> The available NHS exercise machines used in Occupational Therapy or in Physiotherapy are designed for healthy athletes and not for patients with physical problems. The seat of an exercise machine could be developed using car manufacture capabilities to build in a power-assisted function. This could be combined with diagnostic software and optical technologies to visualize the muscle usage and skeletal problems experienced by the patient.[32]

This was an example of co-configuration where automotive designers worked with medical professionals. They were able to develop a new exercise machine by adapting existing products using new technological processes and manufacturing techniques. 'It's about applying engineering design to new medical technologies, and finding out if the patient is working to capacity and how to adjust their therapy'.[33] The redesign of standard equipment was beyond the reach of medical professionals acting alone. In this instance of co-configuration, the 'object of activity' was to combine expertise to help patients with disabilities. The possibilities to create new knowledge were evident in descriptions of individual perspectives that have changed irrevocably. The National Health Hospital Trust perceived the HTC partnership, not as a constraint but as something they would not want to be without.

7.4.4 Vignette: collaboration between defence industry and physicians

The success of the previous project and others like it strengthened the trust between various partners. The NHS, the University Hospital, and small engineering companies sought further created opportunities to collaborate. It paved the way for a larger and wider collaboration, which involved the defence industry working towards an expansive object of activity. The doctors and scientists in the Ministry of Defence had specialized expertise mobile field hospitals.

Practical collaboration in an HTC project changed perceptions about the value of the HTC to local hospitals. Previous collaborations between the NHS, universities, and groups of small engineering companies had built confidence in the capacity to apply their automotive design skills to the manufacture of innovative medical instruments. A more ambitious project, with the researchers in Defence Medicine at DIRA, enabled the medical staff in a large hospital to develop mobile medical equipment, which could be taken out to patients and save them from making unnecessary hospital visits.[34]

> This is about providing mobile medical technologies in the home rather than in hospital. QinetiQ have expertise in setting up mobile field hospitals with electronic data being communicated over great distances. Also we can bring in designers and engineers to develop a mobile kit.[35]

Sustained cooperation blossomed from the growing trust between the public health sector (users) and scientific research communities (developers). Experience in this project created interesting new insights and expanded the horizons and ambitions for further collaborations and on 'expanded innovation partnership projects'. In other words, possibilities for more intensive and ambitious collaborations came to the attention of a major University Hospital.

> QinetiQ at Malvern has the lead in developing imaging systems and is a centre for defence medicine.[36] QinetiQ is not on the Board but we need to make contact between the NHS and QinetiQ and indeed other firms in the UK and have global links.[37]

When people began to see the potential of expanding their horizons, it did not take long before they valued the 'partnership', because it could offer more than the sum of the parts.

> We have been too myopic in the NHS and we should use HTC1 as the means to make that communication happen. MEDICI has encouraged us to train clinicians to be more entrepreneurial.[38] We are much more interested in engagement in partnership working and technology joint ventures have been a catalyst for this. I think HTC1 helped to orientate the NHS to thinking in this direction.[39]

A radical change of perspective seems to have occurred as a result of working with university partners, and recent evidence shows enthusiasm for stronger engagement between three new hospitals and their local universities. This pattern of growing confidence based on step changes in collaborative practices concurs with findings from Engeström's recent studies.

> The occurrence of a fully-fledged expansive cycle is not common, and it typically requires concentrated effort and deliberate interventions. With these reservations in mind, the expansive learning cycle and its embedded actions may be used as a framework for analysing small-scale, innovative learning processes.
>
> (Engeström, 1999: p. 384–385)

Expansive learning is otherwise known as co-configuration, and it shifts the relationships of power and responsibility and relies on the effective communication and accommodation of different perspectives. Although HTC policy designates medical technologies as a high priority for the region, progress is built on small expansive cycles. These build confidence and help to re-orientate local engineering companies towards collaboration with healthcare professionals to co-configure innovative medical instruments and technologies. Successful collaboration in HTC projects creates new opportunities and helps people to reflect on the expertise of partner organizations as shown in Figure 7.4. Projects carried out on a small scale at a local level can create provisional stability. Projects can work in multi-faceted ways as boundary objects and are vital to the 'learning process'. They provide strong foundations from which fragile

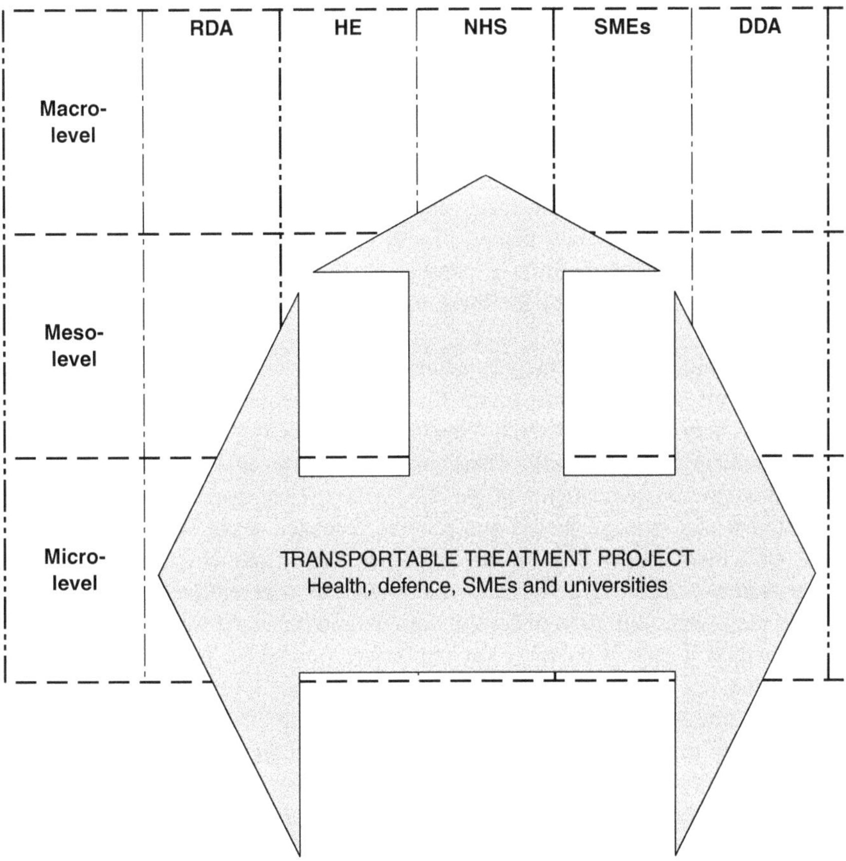

Figure 7.4 Collaboration between hospitals and the defence industry.

and emerging partnerships can grow. Projects can create neutral spaces where boundary mediators can use boundary objects to optimize the potential for diverse organizations to learn together.

This builds a conviction that they can do more together than they can achieve separately.

> Involvement in the High Technology Corridor changed the National Health Trust's **perception of its own expertise**. Before it would have said it was good at everything, but let me give an example. The Leukaemia Centre Programme is to be funded via the RDA – This is a real culture change and very different from previous approaches. It has been a change in practice regarding how we regard excellence [emphasis added].[40]

The NHS hospital trusts and the universities are sharing their research agendas and becoming more focused in specialist fields of medical research and medical technologies as a result. There is a growing awareness of the mutual advantage in synchronizing agendas and effort.

7.4.5 Vignette: embedded projects within Medilink

My interview data showed that a mediator could act as a communication channel, enabling feedback from practitioners at the operational level to inform the strategic policy makers. From the perspective of COPT, Wenger would argue that a mediator is a member of strategic and operational 'COPs'. A mediator needs kudos to facilitate feedback upstream to high-level policy makers.

In the Medilink project, two key actors mediated communications between the policy makers and operational actors. One of the mediators held a senior position in a University Hospital and the other mediator was very influential in a large research intensive University. These mediators were able to play a key role throughout the implementation of the HTC policy. They were able to connect the operational and strategic levels and provide feedback about what was working well. Both mediators were members of the HTC Board and they were able to help other strategic actors to understand 'the impact of concrete practices'. In other words they were able to connect the various contradictory 'objects of activity' that emerged at each of the three key implementation levels. Figure 7.5 illustrates how a mediator can connect people working at different levels.

Nevertheless the role of a mediator is challenging. This is illustrated by the comments of one of these mediators who recognized the complexity of connecting policy and practice and helping people from different organizational cultures to work together constructively in challenging circumstances. Despite this he was consistently positive about the way in which the three local HTC partnerships had held together.

> It hasn't always pointed in the right direction. It hasn't always meant quick action but the commitment of the RDA, of the regional government office and of the major local authorities to the idea, has been an essential component in driving it forward. All sorts of things have got in the way but that commitment has been there all the way through.[41]

At the highest level, cooperation involved the key actors and organizations in shared commitment and passion for the object of activity. The HTCs adopted practices that reflected their commitment to the view that they could do more together than could be achieved separately. This is conceptualized in ELT as their common appreciation of the developmental trajectory of their activity systems (Blackler *et al.*, 2000: p. 293).

The experience of collaborative projects corresponds to the theoretical description of small expansive cycles, which can build fuller expansive cycles of learning. The interview data showed that the partner organizations valued the synergy of working together and planned to do more of this in future.

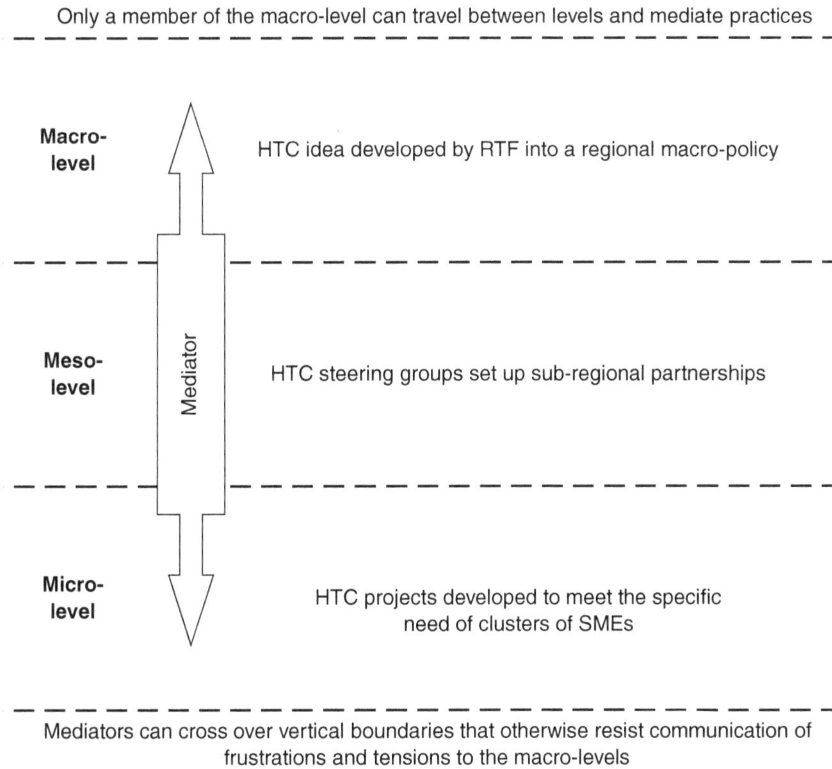

Figure 7.5 How a boundary mediator bridged power difference between levels.

In Scandinavia, healthcare issues have been the subject of recent studies in which ELT helped people to understand processes of innovation and change (Engeström, 2004; Engeström *et al.*, 1999; Tuomi-Grohn, 2003). Studies focused on interactions between activity systems, in which healthcare professionals had previously worked in isolation.

By small steps, these HTC projects created a climate for organizational change. The collective experience from the two previous examples increased the trust and ambition of the partners, which expanded horizons towards the creation of *Medilink*. My interview data showed that Medilink was effective because two powerful mediators acted as champions to connect the resources from the policy makers with a practical delivery mechanism at the project level. Mediators were instrumental in bridging perspectives, one was a senior figure in the university sector and the other was responsible for innovation in an NHS Hospital Trust. Together they linked the needs of a University Hospital and medical research teams with a raft of small engineering companies. Together they had the capability to co-configure new products to solve medical design problems and address new markets. The *Medilink* project entwined several policy strands and integrated a regional *Medilink* network that connected untapped capacity to design innovative medical instruments. It amalgamated new sources of expertise such as in automotive design with new markets in NHS hospitals. This enabled diversification into new 'public sector' markets through the creation of innovative medical products informed by expert groups of users working in collaboration with product developers.

Medilink also worked as a boundary object to connect engineering and medical perspectives towards the 'co-configuration' of new medical instruments as an expanded object of activity.

> In terms of medical technologies, it's about getting doctors talking to inventors. It involves saying what they need from new instruments, based on their experience. It specifies to the technology developers, the information they need to design appropriate instrumentation matched to user needs.[42]

There is an important sense in which 'boundary mediators' can amplify the power of boundary objects by reducing the effect of the dynamics of power in the implementation of the HTC policy that had impaired the development of collaborative practices and expansive learning between levels.

> People who have different approaches to what they do can coordinate their activities around an object, which gives some common meaning across the settings where activities take place. Tensions in regard to the meaning of the boundary object are part of what actors have to take into account in their attempts to coordinate their different interests.
>
> (Tuomi-Grohn and Engeström, 2003: p. 5)

Figure 7.6 illustrates how boundary mediators 'passed through' boundaries *between levels* and how the mediators used the Medilink boundary object help movement *across cultures*. The 'boundary mediators' were able to 'range between' strategic and operational activity systems. In the process of passing through various sites of activity, they change the relationships between people who had previously worked separately. Figure 7.6 shows how the Medilink project attracted resources and aligned the capacity of a vast array of partner organizations.

Figure 7.6 illustrates how the *Medilink* project acted as a more powerful boundary object, when it was used as a tool by 'boundary mediators'.

Medilink created a 'shared discourse', which included ways of talking to one another and ways of 'representing the HTC' to outsiders. Shared discourse tends to form around the concrete achievements of projects, and this can then be used as a tool to link the perspectives of different organizations. Another example of this was the collaborative creation of an HTC Web site discussed in vignette 'collaborative projects to develop electronic tools' in Chapter 5.

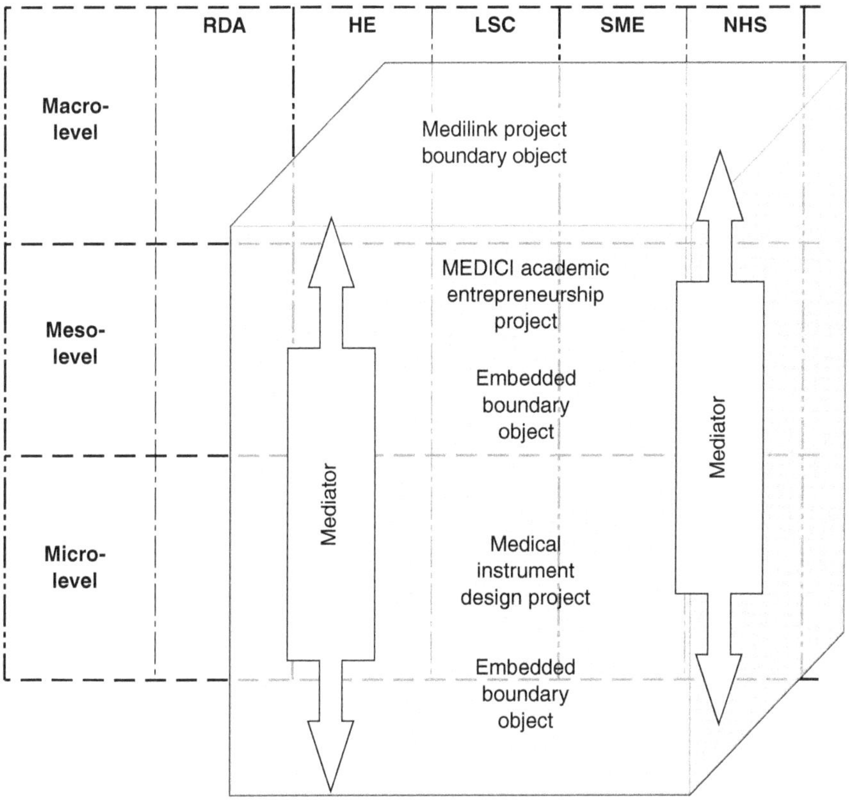

Figure 7.6 The power of an embedded boundary object used by boundary mediators.

7.5 Overview

The contrast between the structures and processes for innovation at Google, at Nuno, and in the HTCs is remarkable. Complexity and hierarchical power structures create obstacles to efficient innovation. Google enables self-organizing project teams to innovate and these projects are peer reviewed, so the decision to commercialize is not determined by the senior management. Nuno adopts a highly focused approach to innovation and selects partners with great care. The success of a project depends on the skill of a mediator who understands the practices of the various industrial sectors and partner companies. In both Google and Nuno, experimentation is encouraged and the risk of failure is accepted as a natural consequence of the innovation process.

The disparity between innovation driven by public sector funding and the approach to innovation adopted by successful private corporations is starkly illustrated by the HTC vignettes. These showed that complex innovations and lengthy implementation chains made organizational learning extremely challenging. It is clear that small focused projects are flexible and effective vehicles for innovation, whereas it is very difficult for complex multi-agency partnerships to deliver innovation per se.

Despite these difficulties, the vignettes showed that as partners learned how to work together in the field of medical technologies, they were able to co-configure new products and services because they were all working towards an expanded 'object of activity'. Expansive learning was an evolutionary process. Confidence in the *Medilink* partnership grew from the development communication strategies to publicize the achievements of individual projects and promote the future plans. At the same time, the discourse engendered by *Medilink* became a tool and facilitated more expansive and ambitious collaborative projects. *Medilink* used both boundary mediators and boundary objects to create very effective bridges between horizontally and vertically arrayed activity systems. Eventually, this resulted in sustained collaboration between the partners and organizational learning. I will now turn to Chapter 8, which focuses on evaluation as a tool for organizational reflection and learning.

8 Using evaluation as a tool for learning

8.1 Variations in the style and purpose of evaluation

The rhetoric of *evidence-based practice* is prevalent in national and international policy documents on innovation, healthcare, and education. Evidence is derived from an evaluation of new practices, and ideally, this involves a virtuous cycle of learning whereby the trajectory of policy can be adjusted to respond to the known effects of new practices. However, *causal knowledge* will often be uncertain or even unavailable, and it is important to consider strategies to cover a wide range of evaluation perspectives (Bredgaard *et al.*, 2003).

Many books on evaluation discuss the effects of political contamination of evaluation. Vedung's (1997) *Public Policy and Program Evaluation* argues that the political environment is the necessary foundation for public policy and program evaluation. Vedung proposes a model that incorporates decision-makers' goals with 'side effects' because interventions often yield unintended consequences. He recognizes that theory and practice can be mutually illuminative and include opportunities for practitioners 'to learn by being involved in evaluation'. This can lead to the development of new theories informed by reflective practice.

The effects of dynamic political processes involved in implementing complex policies and programmes are unpredictable, and so feedback from practitioners can help policy makers to understand what is going on. The people involved in implementing a policy can learn more from their activities if they are involved with deconstructing the programme's logic with an evaluator (Valovirta, 2005). This can help to clarify the perception of the changing purpose of the project and the overall programme as the partnership evolves.

> There are many factors, which make the identification of programme logic very difficult, like fuzzy borders of the programme, non-linear causality context and multitude of relevant implementing actors … . However, whenever it is possible, the cause-effect relationships should be elaborated in more detail, by linking the programme theory with investigations on the mechanisms, which drive the change process.
>
> (Valovirta, 2005)

Christiansen (2001) suggests that policy makers often change their minds about how to achieve more efficient innovation because 'oscillating public priorities, systemic interdependencies and institutional inertia may hamper the processes of technological change and innovation'. Stame recommends a better interaction between lower and higher levels of the hierarchy. In this way, there would be more cross-fertilization between theory and practice, so that good practices would be the subject of reflection and would influence the formation of good theories. In other words, it is valuable for people to link organizational learning with the construction of theories based on experience.

> But nothing of the kind is possible inside the existing institutional hierarchy of the EU system of evaluation, where
> - lower levels are expected to do only monitoring, not real evaluation
> - higher levels do not receive contributions from below.
> - at the higher level, theories received from below would be more grounded, and better suited to understand a complex and diverse situation.
>
> (Stame, 2003: p. 10)

However, interventions with public funding by a government involves bureaucracy, and there is evidence that even a partial investment of public funding slows commercialization and constrains risk taking. Aversion to the risk of failure inhibits progress since innovation depends upon developing new ways of doing things and 'change involves loss, anxiety and struggle' (Fullan, 1993: p. 30).

In Europe, the reluctance of the private sector to fund innovation is a market failure that contrasts strongly with the rapid commercialization processes that enable the WSA to forge ahead. A burdensome bureaucracy is likely to impede creativity, change, experimentation, and the tolerance of risk. A further difficulty arises because many EU policies on innovation are actually dominated by social cohesion objectives. The outputs are jobs created or protected rather than wealth creation or 'return on investment' (ROI). Recent research shows that effective evaluation can follow three steps:

- Step 1 determining the policy objectives against which to assess its effects;
- Step 2 finding the effects of a policy;
- Step 3 combining effects due to the policy with the objectives in order to determine effectiveness.

> To carry out any of the three steps involves many challenges. These include determining which effects are caused by the policy being evaluated and to what extent, whose goals should be taken into account and how to interpret the combination of objectives and effects.
>
> (Mickwitz, 2003)

The analysis of the case studies suggests that each unique situation can benefit from a cross-pollination and sharing and adaptation of strategies that have been successful elsewhere, but they must not serve as blueprints.

> Organizations' ability to change their ways of working depend deeply upon their histories, their persistent institutional characteristics, and upon the complex interactions among their many actors … . Nevertheless, organizational innovations can be significant and can ultimately be copied by others (at least in part).
>
> (Fine and Raff, 2001)

Enduring changes in behaviour are built slowly from experience and trust, and consequently, policy that attempts to import a 'culture of innovation' is flawed. It is evident that cultural change cannot be brought about by 'policy borrowing'. It is unlikely that 'processes' effective in one context can be successfully transplanted into another. '[T]he transfer of procedures involves a number of contingencies that cannot be planned in advance. Procedures, because of their complexity, must be transferred on an ad-hoc basis' (Rebentisch, 1994). This view is supported by international critiques, which alert policy makers to the risks of policy borrowing since sustainable cultural change arises from seeing the benefits of adopting new practices. 'a) nothing can be considered best for all situations, hence generalizable b) there are different situations, and something that has shown to be good somewhere perhaps could be studied and adapted/imitated somewhere else' (Stame, 2003: p. 9). The development of 'new ways of doing things' needs to grow from shared experience of working things out together.

8.2 How involvement in the process of evaluation supports learning

There are risks and possible misuses of evaluations. For instance, evaluations can be used as a policy justification tool rather than as a mechanism for shared learning and improvement of both policy formation and implementation. However, evaluation can be used constructively, as a powerful tool to help people to analyse a problem systematically and reflectively.

> Ideally the evaluation should be to activate cooperation to find out promising practices and procedures for networking and partnership. Often evaluators face evaluation tasks where the principal endeavour turns out to be figuring out what the programme really is that should be evaluated. The programme logic or programme theory needs to be clarified first, only after which the programme can be sensibly evaluated. In fact, the programme intervention logic often has to be constructed, since it has not been explicitly expressed by.
>
> (Valovirta, 2005)

Vedung's approach to evaluation aims to bring clarity to a contested and politically sensitized process. It also begs many questions, which it may be difficult for policy makers and the commissioners of evaluations to answer with complete honesty. If practitioners are involved in the process of evaluation, it is possible to focus time and effort on developing human networks and communications and to adjust the policy and the budget towards achieving this end. Innovation flourishes in a culture of trust, where people understand and accept that the huge potential benefits are coupled with the risk of failure. However, as indicated in the vignettes in Chapters 6 and 7, a culture of control, accountability, and auditing tends to have a negative impact. These factors also inhibit meaningful evaluation as indicated in Table 8.1.

The next section illustrates the need for evaluation by reference to stories of contradictions and dilemmas experienced by the individuals responsible for implementing the high-technology corridor (HTC) policy in the UK. Reconstructions of 'experience' from the interview data refer to contradictions between the espoused policy process and what happened in practice. However, the creative power of such contradictions to drive expansive learning is realized only if the contradictions are acknowledged. The following sections explain why recognizing contradictions and reflecting on problems is difficult for the policy makers.

8.2.1 Vignette: the victory narrative and reflective evaluation

There is an inherent contradiction between recounting a success story and learning from what went wrong. While a victory narrative tends to justify the time, effort, and cost of the HTCs, it inhibits the process of reflecting on mistakes and 'learning from practice'. Evaluation was not perceived as an opportunity for learning. A culture of anxiety and blame seemed to have contributed to a reluctance to evaluate and reflect on what happened in order to inform future plans (Vince and Saleem, 2004). 'Lack of collective reflection undermines the practice of communication between hierarchical layers and across the boundaries of organizational sub-systems' (Vince and Saleem, 2004: p. 145). In 2005–2006, these insights were shared with key informants. A key actor in the HTC programme strongly agreed with this 'Victory Narrative' analysis and made the following comment:

> The RDA had never engaged in any meaningful evaluation and have merely carried out measurement observation.... They were frozen (like rabbits in headlights) and unable to be able to take risks or be seen to make mistakes. They were unable to acknowledge the experimental nature of HTC.[1]

He also strongly agreed with the diagnosis of a culture of anxiety and blame, which had contributed to the unwillingness to reflect on practice. The capacity to reflect on experience is enriched by metaphors because of their capability to visualize juxtapositions between the different elements of that experience. The creative

Table 8.1 Key issues in evaluation

The question	*What needs to be studied*	*Examples from the case studies*
1 The purpose problem	For what overall aims is the evaluation launched?	Is it used to justify the expenditure of public funds? Does this preclude learning from what may have failed? See vignette 8.2.1
2 The organization (evaluator) problem	Who should exercise the evaluation and how should it be organized?	Who funds and briefs the evaluator? Is this a quick and dirty victory narrative? See vignette 8.2.1
3 The intervention analysis problem	How is the evaluand (e.g. the policy or the provision of services and goods) to be characterized and described? Is the evaluand regarded as a means or as a self-contained entity?	In the terminology of activity theory 'Are we evaluating a tool or an outcome of activity?' Vignette 5.3.5 illustrates the transformation of one element of activity into another during cycles of expansive learning
4 The conversion problem	What does execution look like between the formal instigation of the intervention and the final outputs?	How was the policy implemented? Was there a gap between the policy (theory) and what happened (practice)? See vignette 8.2.4
5 The results problem	What are the outputs and the outcomes – immediate, intermediate, and ultimate – of the intervention?	How were these various outcomes understood? Vignettes 7.4.4 and 7.4.5 illustrate small cycles of expansive learning lead to fuller more extensive collaboration or expansive learning
6 The impact problem	What contingencies (causal factors and operating causal forces) – the intervention included – explain the results?	What was the impact? Did the policy achieve its desired effects? What were the contextual factors?
7 The criterion problem	By what value criteria should the merits of the intervention be assessed? By what standards of performance on the value criteria can success or failure or satisfactory performance be judged? And what are the actual merits of the intervention?	Was there an agreement on the success criteria? Did the policy have any detrimental effects? Conversely, were unforeseen benefits attributable to the policy? Compare vignettes 7.4.2 and 7.1.2. Google has a very clear innovation policy and success criteria, but the complexity of HTCs makes clarity difficult
8 The utilization problem	How is the evaluation to be utilized? How is it actually used?	What is to be done as a result of the evaluation? How are lessons to be shared and policy adjusted? See vignette 8.2.4

Source: *Vedung's eight-problem approach to evaluation* (Vedung, 1997: pp. 93–94).

power of metaphors frees up our capacity to imagine new ways of doing things in the future. Some metaphors spoke indirectly of problems, which tended to be difficult to articulate directly. 'Blame undermines the ability of managers to reflect. Lack of collective reflection undermines the practice of communication between hierarchical layers and across the boundaries of organizational subsystems' (Vince and Saleem, 2004: p. 145–146). Evaluations of complex programmes such as the HTCs are complicated because of horizontal and vertical complexities. It cannot be assumed that when activity systems are of equal agency and arrayed in parallel, they will interact in the same way as activity systems arrayed vertically where this is a distinct pecking order. The impact of evolving contexts and the breadth of potential outcomes make interpretation problematic. Political, demographic, and geographic factors indicate the unprecedented complications of the task (Connell *et al.*, 1995: p. 2). There were no published studies of comparable research contexts or was it possible to use a control group for comparison purposes.

> Politics can inhibit theorising. It can limit the scope to concentrate on issues and options that fit the current political agenda and fail to articulate and test alternative sets of assumptions-or alternative causal stories – in which case we are likely to 'know more but understand less'.
>
> (Weiss, 1995: p. 11)

It is possible that reluctance to evaluate stemmed from a realization that there was little opportunity to make any meaningful changes because all the money had been profiled. However, a full account of the development of the HTCs, with reference to the difficulties as well as the achievements could be an invaluable resource for policy makers and for the people implementing it in practice.

This perspective seems to be supported by the UK case studies in which changeable and volatile priorities were detrimental to the implementation of innovation policy. Political factors were associated with a highly visible expenditure of public funds. The public attention to the potential redundancies at Rover skewed the role and power of different agencies within the partnership. There is a risk that attempts to specify how relationships and operational activities should happen can stultify the people and organizations concerned.

> a project-management model is a socializing force capable of defining, aligning and disaligning heterogeneous actor groups, work tasks, relations and responsibilities. It is a means by which discourse technologists can redesign work processes, turning them into norms, rules and prescriptions, representing established consensual praxis.
>
> (Raisanen and Linde, 2004)

This is not only dysfunctional to learning, it is also self-deceiving, in that it encourages negative behaviours such as acquiescence, resentment, and fear of expressing one's true opinion. The resulting power struggle changed the 'object of activity' from innovation towards 'politically sensitive' issues and prioritized stemming job losses, which could be easily measured in ways that met the EU funding guidelines. The original intention of stimulating innovation in the supply chain was undervalued because the benefits would be only seen in the long term. 'Engineered culture is oppressive – because it systematically encourages a sense of individualism that devalues genuine respect for other people' (Ezzy, 2001). The relationship between the private and public sector was put under intolerable strain by the audit cultures imposed by a raft of public finding mechanisms. Opportunities to change habitual behaviour and to promote positive attitudes towards innovation resulting in cultural change were almost missed entirely. The situation was rescued only by the dedication of a few enlightened and powerful mediators who fortunately were able to influence those in control. 'Organizational innovations can have significant consequences for economic performance. But such innovations appear to be much more difficult to import from one company to another than product innovations' (Fine and Raff, 2001). In attempting to take a short cut to achieving innovation through organizational partnership, there is a risk that importing behavioural patterns does not have the desired effect. It is over simplistic to assume that it is possible to transplant a behaviour or a culture from one context to another. As policy becomes more complex it is even more important to learn from a process of on-going evaluation, so that evaluation becomes a tool to adjust the trajectory of the policy to attain the desired effects.

A recent evaluation of cluster development across the nine English regions sponsored by the Department of Trade and Industry illustrates the insights a detailed evaluation can bring to a complex policy initiative (Ecotec and DTI, 2004). The DTI evaluation adopted a sophisticated approach to the problem of technology transfer that recognized the complexity of interactions and the impact of social networks. Whereas previous DTI reports focused on vertical links and supply chains, 'clustering' recognizes that it may be more important for companies to have horizontal links through complementary products and services and the use of similar specialized inputs, technologies, or institutions.

'Most of these linkages involve social relationships or networks that produce benefits for the firms involved' (Ecotec and DTI, 2004: pp. 4–5). The DTI evaluation was based on interviews with practitioners across the UK, many of whom had learned a great deal. The 'lessons' presented in the evaluation of 'clustering behaviour' recognize the complexity of organizational relationships. However the HTC experiment was initiated much earlier and these lessons were not available to the Rover Task Force in 1999 when the implementation of HTC policy began. At that time the importance of evaluating the implementation of policy was not understood. Indeed in 1999 the problems facing Rover were unprecedented in the UK.

> The aim of measuring cluster development is so that we can learn from our experience in order to improve the actions that we are taking. Unsuccessful actions should be ended, successful actions continued, replicated where

> relevant or discontinued where no longer appropriate. New actions should be adopted where information suggests that weaknesses are emerging or opportunities present and things should be left well alone where no actions are needed. Monitoring and evaluation is part of the policy cycle and information gained from measuring cluster development should be fed back into the policy process to inform future policy development. This requirement may itself influence the timeframe adopted for the monitoring of cluster development.
>
> (Ecotec and DTI, 2004: p. 19)

It is remarkable to note the change in the vocabulary and tone. Some reference is made to considerations central to sociocultural perspectives and which have hitherto been missing from the various theoretical concepts that inform policy makers in the UK. The report focuses on 'clustering behaviour' and the experiences described by practitioners as a resource for mid-cycle policy adjustment. Figure 8.1 illustrates this change in terminology and attitude.

The report seems to value both social networks and individual agents of change, and there is some use of terminology that has not previously been evident in UK Government-sponsored reports.

> Networks that generate formal and informal flows of knowledge and information throughout a cluster provide the gel that binds success over time. Access to **tacit knowledge** can support **collective learning** and more competitive performance. The value of informal networks, based on social relations and even job-moves, is that it enables a transfer of knowledge around the cluster. Such 'untraded' means of information dissemination such as informal collaboration and extensive contact networks can create a '**knowledge community** [emphasis added]'.

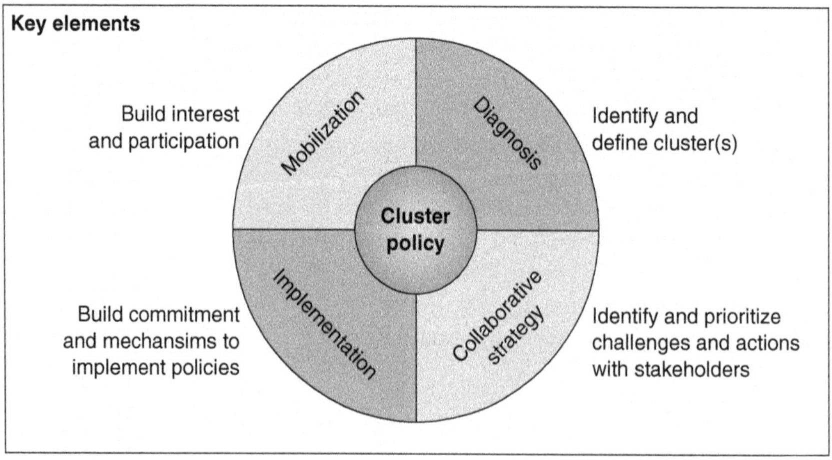

Figure 8.1 Guidelines on clustering, 2004.

The evaluation report refers to clustering activities and to a 'knowledge community' rather than to a knowledge economy'. Given that the report was commissioned by the DTI, the shift in language is remarkable. It describes the process of clustering in terms not dissimilar to those used in socio-cultural theories, although this is clearly one of several influences on the authors.[2]

> Cluster networks need to be more than simply opportunities to meet. Networks will ideally form '**communities of practice**', with many such networks present in each cluster, associated with different interests. What causes one company or interest to join a cluster will not be the same for everyone and their needs are likely to shift in time. It is likely that networks will continuously form and reform as membership and needs change [emphasis added].
>
> (Ecotec and DTI, 2004: p. 24)

Despite the espoused mission 'to facilitate regional innovation', the 'organizational culture' and performance measures constrained the capacity for reflection, self-criticism, and expansive change. By recognizing inadequacies in current practice, it is possible to identify better ways of doing things and drive progress. The difficulty (or to use Engeström's term, the double bind) of public sector organizations is the dilemma of 'being seen to make mistakes'. The cumulative 'weight' of the bureaucracy tended to deter policy change on the basis of sustained reflection, local knowledge, and operational experience.

The ability to bring controversial issues to the 'surface' is a prerequisite for reflection. There is an understandable reluctance 'to be seen to have funded' a failed experiment with public money (Vince and Saleem, 2004). There was no evidence of attempts to develop consensual theories of change. Similarly, there were no opportunities for evaluators to help people to work together by developing techniques to measure the extent to which each step has taken place or to consider: 'Have agencies adapted their procedures in ways that enable them to function in a multi agency system?' (Weiss, 1995: p. 11) and 'Have practitioners reinterpreted their role to be advocates for the client, rather than enforcers of agency rules?' (Weiss, 1995: p. 11). The empirical data suggest that Regional Development Agency (RDA) staff did not reinterpret their role to be *advocates for the client*, instead they remained focused on enforcing the agency rules. There did not seem to be any systematic method of evaluating progress and trying to comprehend how barriers might be overcome. In 2006, a key actor in the HTC programme commented on this analysis and emphasized that the RDA were incapable of recognizing the need to treat HTCs as *high-risk experiments*. The agency concentrated on enforcing rules and would and could not recognize that this became part of the problem. 'The RDA were incapable of recognizing the need to treat HTCs as high risk experiments. The agency concentrated on enforcing rules and would and could not recognize that this became part of the problem'.[3] RDA staff focused on making sure spending profiles were accurate and exhibited 'risk-averse behaviour' which stemmed from a culture in which accountability and 'auditability' usurped the intended outcomes of HTC policy in importance. The corollary of this was 'much frustration' at the meso- and micro-levels and a waste of human and financial resource.

In target-driven or project management-driven cultures, it is very difficult to acknowledge problems and to reflect on current practices. HTCs may have fared better if they had been recognized as 'high-risk' experiments, which carried potentially high rewards for success. A positive attitude towards risk may have encouraged reflection and adjustment to policy, informed by local insights. The logic of embedding an element of qualitative analysis and evaluation from the outset seems compelling since it allows not only for a rich narrative but also for modification of causal assumptions as things happen in the field (Weiss, 1995: p. 11).

> When we come to the longer term policy, we hope not to have just a series of catalyst projects which get done, and then the thing dies and they say 'Fine we've had some money for projects now we move on to something else'. Unless the corridor has got a long-term aspiration then it will have **failed** [emphasis added].[4]

The HTCs continue to be an interesting experiment in which empirical and theoretical knowledge could helpfully reconstruct one another. Expansive learning (or collaboration) seems to progress when there is an open discussion of difficult issues. The lessons learned in this ambitious and 'high-stakes' policy initiative would be relevant and transferable to similar technology transfer initiatives elsewhere, but only if a complete story is told. This would include the difficulties and tensions that arose in the implementation process.

> My impression would be that its like all the partnerships we set up, that the RDA got going, that it takes time to bed down and it takes time for people to realize what their roles were, what was expected of them, what they contribute and they are now starting to become much more effective.[5]

The requirement for a victory narrative, focusing only on a success story with no acknowledgement of setbacks and deviations to avoid pitfalls, limits the capacity for real change. It is not very useful to people struggling with practical problems of a similar nature, whereas a frank and full exposition of the story of HTCs from multiple perspectives could be invaluable.

8.2.2 Vignette: sustaining coherent development with short-term projects

The contradictions between the short-term regional 'objectives' of the corridors and their long-term survival in a global economy caused dilemmas for the HTC steering groups.[6] Although the stakeholder partners were aware of the need to plan for the future, their freedom of action was constrained by the terms and conditions of the funding from the DTI and the EU. Many regional problems were deeply entrenched and required sustained and consistent treatment over decades. Two- or three-year projects simply disguised the problem.

There was a tension between the need for a long-term strategy and the fact that public funding was available for only the first three years of operation.[7] The complexity of the investment and revenue funding streams for the HTCs caused huge practical difficulties, which gradually became apparent during the implementation process. The dilemma at the meso-level facing the HTC directors and their steering groups resulted from the tension between short-term targets and long-term sustainability.

> We have funding for only three years but clustering takes 15 or 20 years to achieve. We need to prolong the project and expand it from the geographical region defined by the HTC3. I think clustering is best done internationally and we are discussing working with companies across the UK and then we want to expand worldwide. In particular we are keen to link with Indian companies. Local funding conditions make this forward vision difficult. The RDA want us to focus on achieving local targets.[8]

Contradictory and inconsistent policy messages were confusing, because on the one hand, the HTC funding was time limited, but on the other, the problem the policy sought to address was long term. Sometimes, the officers at the RDA conflated the two issues and used emotive language in describing the HTC as a long-term solution. 'Unless the corridor has got a long-term aspiration then it will have **failed** [emphasis added]'.[9] The key actors in each HTC described the process of persuading the RDA to see the contradictions between short-term funding and long-term aims as *a long uphill struggle*. While the focus has been on delivering the polymer cluster and Advanced Engineering projects in the HTC3, sustaining these projects to meet a long-term need was also considered.

> While I think the policies have been supported and the corridors are now seen by the RDA as one of their long-term strategic regeneration vehicles and have begun to talk about a 20–25 year lifetime. Getting that expressed consistently by funding strategies is always a little more difficult. I think that where the difficulties have arisen is where the rules around matched funding and spending profiles created confusion around **output targets**. So I think that the policy framework is supported but **the practical interpretation** of that through funding is more difficult [emphasis original].[10]

After a good deal of reiteration of the problems experienced at project level, changes were made to the 'targets and performance measures' which were applied to some of the partner organizations working in the HTCs. During 2003, Business Link and HTC actions were aligned to alleviate the impact of performance measures to reduce self-interest and help to align the partners towards the same goal. The efforts at the micro-level to influence the macro-level eventually resulted in a change in 2004, when the RDA adopted new strategies to encourage closer working relations between local public bodies.

Individual officers within the RDA were aware of the difficulties but were powerless to ‘change the rules’. A complicated chain of approvals was required before anything practical could be done at the micro-level, and this proved very frustrating for the SMEs, who were the espoused beneficiaries of the HTC policy.[11]

> Given that we didn’t have well founded partnerships in the early days at least it was important for external agencies to take a very strong view on whether or not things were approvable and in the case of the ones where we had RTF money – they had got a plan to work to and any things that were not in the plan were not going to get signed off by the Secretary of State.[12]

As a result of persistent lobbying from the HTCs about their approach to managing this diverse array of public funding, their perspective shifted during 2003 to allow a more flexible approach to address issues of sustainability.

> A more positive indicator is that the RDA are now talking about continuing the concept to 2010 and plan to include references to corridors in the RES (Regional Economic Strategy). It may be that the RDA now appreciate that a 10–15 year cycle is required to assure the desired payoff for the investment.[13]

This realization came too late to redress the balance between capital and revenue expenditure. In effect, no additional resource could be made available to address the problem of sustainability. Unfortunately, as in other instances, the messages from the EU funding regimes and from the RDA were contradictory and confusing. Ironically, this contradiction was a driving force for expansive learning in the meso-level, and collaboration between the members of the HTC steering group became deeper. ‘The RDA did not become more flexible; they merely shifted their position to allow a more sustainable approach’.[14] In a sense, the partner organizations began to unite against the common enemy of ‘bureaucracy’. Their object of activity was *re-orientated* towards overcoming obstacles imposed at the macro-level. The research findings suggest that it would have been valuable to realign existing partners, to re-conceptualize the problem, and to evaluate progress and problems together.

> I think that where the difficulties have arisen is where the rules around matched funding and spending profiles created confusion around output targets. So I think that the policy framework is supported but the practical interpretation of that through funding is more difficult.[15]

During policy enactment, it gradually became evident that there were huge problems associated with accepting a funding package of extreme complexity with rigid spending profiles.

> We've tried along the way, for 2 or 3 years I suppose, to keep industry and commerce in the corridor briefed about what was happening and about the opportunities. It was quite difficult. I mean the progress for the past 3 years has been about progress with bureaucratic procedures about money and that's seemed pretty difficult to comprehend I think unless you have been in the middle of the process. **It's sometimes difficult to comprehend when you've been there** [emphasis in original].[16]

This resulted in almost no flexibility to address surprises and adapt practices. The various terms and conditions from elements in the complex funding package were contradictory and confusing. Overall, this contributed to a quick fix, 'project' culture in which numerous short-term projects proliferated to solve intractable economic problems. 'To do economic development properly you need a much longer-term vision. So although we've got a vision you need longer-term resources to have a real impact'.[17] Different types of innovation have specific effects. A *solution innovation* typically applies only to the specific case for which it was invented, and often, this may not have a lasting or wider effect. On the other hand, a *trajectory innovation* is aimed at becoming a more or less permanent repeatedly used procedure (Fichtner, 1999: p. 387). Although *solution innovations* can sometimes be transformed into *trajectory innovations*, the process is difficult to achieve in a climate of short-term 'project funding'. 'When an innovation is rejected but has conceptual coherence and is anchored in the needs of the collective activity system, it is not likely to disappear, but subsequently may reappear in modified forms' (Engeström, 1999: p. 397). Sustainable *trajectory innovation* is difficult to achieve in a climate of short-term 'project funding'. In other words, it was unrealistic to expect to sustain innovation with discrete short-term interventions. Furthermore, the cost of sustaining a consistent service should have been compared with the opportunity cost of failing to broker or mediate communications between users and developer communities. Perhaps this lack of understanding was because the English RDAs are relatively new organizations. Compared with economic development agencies in other European companies, they seem unstable and transitory.

> Italian development agencies are **stable** over long periods and are not subject to rapid policy change determined by changes of Government. (In Italy the Governments change very frequently, and so stability for regional economic development is crucial.) SMEs need longevity and continuity in the UK. The phrase **'please stop helping me'** sums up the feelings of SMEs about the confusing array of short-term support available in the West Midlands [emphasis in original].[18]

In many instances, new regional initiatives were launched prematurely before a prior 'solution to the problem' had played out to its conclusion. *Initiativitus* arose as the short-term funding of a 'solution innovation' reached an end (Brown, 2004). Instead of being transformed into a sustained trajectory innovation, the same good idea re-surfaced time and again as a re-branded solution innovation. Not surprisingly, this caused confusion among the very people it set out to help.

This risks repeating local policy failure, because sustained cooperation between complementary roles of regional agencies was disrupted in the hiatus between one short-term solution and the next. The cost of sustaining a consistent service was not compared with the 'opportunity cost' of failing to broker a sustained dialogue between users and developer communities, without which innovation would not succeed.

The HTC policy fuelled unreasonable expectation that economic and social problems can be solved quickly. An exclusive focus on hard data assumes that 'what is measurable is most important'. The soft data pertaining to cultural change in SMEs were not valued because it did not correspond to concrete outputs such as the number of new companies created or jobs protected. Ironically, the hard outputs may not be appropriate to measuring performance in the new global knowledge economy.

During 2004, the empirical data showed that the perspective of the RDA shifted to allow a more flexible approach, which addresses issues of sustainability. 'I think there was a realization that business support needs were specialised – a different mix was needed – The planning framework is not set up to acknowledge this particular type of problem'.[19] In the financial year 2004–2005, the RDA took responsibility for monitoring 1,800 projects.[20] RDA personnel were accountable for the delivery of outputs on all these projects to the UK Government and the EU. In anticipation of some difficulty, efforts had been made during the previous financial year to cut this down to manageable numbers and to focus on ten major regional projects. However, it proved to be impossible to jettison existing projects or to rationalize them by aggregation.

8.2.3 Vignette: conflict between the roles of animateur and auditor

Difficulties arose for all the English RDAs because of the ad hoc development of policies in response to very tight funding deadlines driven by the UK Government's economic performance targets. In the West Midlands, a local 'critical incident', known as the Rover Crisis,[21] applied additional pressure to the situation. In effect, there was a *post hoc rationalization* of a suite of technology transfer and innovation policies devised in haste. These now appear to be full of tensions and contradictions, which need to be understood, in order to change future practice. Internal contradictions between two different cultures at the RDA contributed to miscommunication, wasted time, and effort. The RDA acted as both a regional animateur and the 'gatekeeper and auditor' of public money.

> This tendency is getting worse, and the i54 (a science park joint venture) is in jeopardy because the RDA over-committed their revenue funding and are willing to sell capital assets (such as the i54 land) to the highest private sector bidder, regardless of their suitability, just to show a return on investment.[22]

There were inherent tensions in these two roles. In their dual role, the RDA did not recognize this *dilemma* and their internal contradictions of organizational purpose created problems and confusion for the so-called 'partner' organizations. For instance, the comments of a member of an HTC steering group can be related to a depiction in activity theory of internal conflicts and 'institutional schizophrenia' within the RDA.[23]

> Because it does not fit into a standard contract model some of the RDA staff are reluctant to move things forward. At the root of the problem is a dilemma as to whether the RDA is a **delivery organization** or a '**strategic planning body**'. There are almost **two cultures** at work in the RDA pulling in different directions. But there is not the capacity to deliver even though there may be the ambition to do so [emphasis in original].[24]

Two cultures existed side by side at the RDA. This may be summed up as *driving with one foot on the brake and the other on the accelerator*. Respondents at all levels and from several of the stakeholder partners inferred that the RDA gave mixed messages and exhibited contradictory behaviours. 'The issue is about the **funding bureaucracy** being at odds with the timescales for project delivery. It makes it difficult to plan "forwards" in a consistent way [emphasis in original]'.[25] One team of people applied strict rules from an audit perspective, and this contradicted the efforts of a different team of people whose role was to enthuse regional organizations to join the HTC policy partnership. 'The partners are disparate and (getting agreement is slow). The RDA is one of the partners but their funding mechanisms slow things down'.[26] Tensions within and between the activity systems drive change. It may bring people closer together or drive them apart. An unintended and fortuitous result of the frustrations arising from the complexity of the rules of the various funding regimes was that the HTC partners drew closer together. The tension between the different sets of rules and 'competing objects of activity' in the HTC united the partners. A 'wartime spirit' prevailed, and they agreed unanimously to fight the inflexibility of the rules.

> Because we had such a tough time sorting out the HTC funding vehicle (state aid rules were a big problem), I think we started concentrating on land and site acquisition. The revenue expenditure for spinouts has taken a back seat.[27]

In HTC1, the 'resistance' to bureaucratic funding approval processes resulted in a decision by the steering group, to gain more control by changing the legal status and name of their HTC.[28] Consequently, the steering group members established as a company limited by guarantee and adopted a new name and set a twenty-year strategy. They asserted their independence from the RDA and changed the dynamics of relationships. Their action prevented the RDA from dominating the partnership by virtue of their status and power as *gatekeepers* of funding. Nevertheless, the RDA continued to give mixed messages to the so-called 'partner organizations'.

> They had to own the corridors-they are **not** owned by the RDA.[29] The RDA is the **funding route**. They have to develop their policies and get those individual projects up and running and establish their long-term vision and their long-term plan. So that really is the implementation and that really is the stage where we are at the moment [emphasis in original].[30]

Tensions continued to arise from 'conflicting rules' associated with the Rover Task Force funding package. There were persistent difficulties that took a great deal of patience and energy to resolve. The complexity of the HTC funding package encouraged the use of measures such as jobs created on all projects. In some instances, this made no sense because in the Information Technology or Design Industry, freelance employment conditions do not make the number of jobs created a reasonable measure of success. 'Unfortunately many of the old performance measures that apply are support measures, which are out of date and inappropriate to what is needed'.[31] For instance, profitability would reflect more accurately the improvement in business performance, but the EU or UK Government funding programmes will not accept it as a performance measure. The overall impact of utilizing a mixture of European Social Fund (ESF), European Regional Development Fund (ERDF), and DTI money was not appreciated until there was no turning back.

> We were pretty much constrained by what went into the Secretary of State . .. it didn't incorporate some of the things we now know needed to have been included (and we are going to find the funding from elsewhere to make them happen). It's more than a package of knowledge intensive activities and land and property propositions.[32]

Interviews with key actors illustrate tensions and contradictions between how the policy is portrayed and how it is experienced. In the same interview in 2003, the lead policy officer voiced some revealing self-contradictions.

> The agency owns in effect the corridors and we have funded the people in the corridors who are administering the running of the corridors. In the case of the HTC1 there is a Management Board. In the case of HTC3 and HTC2 we're funding a corridor Director, and although we're funding them, and they report to the group, effectively they report to us.[33]

Another consequence of the rigorous auditing of individual projects was the sheer effort and staff time this absorbed. This tension was embodied in a contradiction between the espoused autonomy of the HTCs and the control mechanisms imposed by the RDA's project management practices.

> So we're monitoring the longer-term policy development of the corridors so there is that level and then as far as the individual projects are doing, we are funding and monitoring these as well. So in effect there are two levels of monitoring.[34]

Within the same interview, an officer from the RDA said exactly the opposite and stressed that each HTC was responsible for 'developing' policy. However, the senior manager (and line manager) never wavered in the opinion that 'the RDA owned the corridors'.[35] It was implied that paying the salaries of the directors, equated to owning the corridor agenda. There were inherent contradictions between the espoused autonomy for the HTC steering groups, which included creation of a long-term vision, and the treatment of each project by the RDA as if it were independent from the HTC plan.

> We've now funded corridor Directors, so from a corridor perspective they start bringing everything together and are able to give us a **report** on the delivery in terms of quality. The fact is that the expenditure is being managed and the outputs are being managed, so Directors are employed by the corridor **but** report to us (the RDA) in a **contractual** sense [emphasis in original].[36]

There was no recognition of inconsistency or internal contradiction in the detailed monitoring requirements at project level within each corridor, and RDA priorities were clearly to ensure that public money had been spent correctly and had a robust audit trail.

> Until further delegation happens every single project in the corridor is on a contract basis in terms of spend and outputs on a monthly basis. The reports are on a quarterly basis and that's followed up by on-site visits so it would be standard agency monitoring really.[37]

The HTC policy rhetoric constantly spoke of the need to support the business community and listen to what they wanted. In practice, the business community was very confused by a vast array of projects with different eligibility criteria and funding rules. Their needs were not well served during this period. 'The RDA have not become confident enough to allow the local people on the HTC steering groups to "own the corridors" and this attitude has been disastrous'.[38] The *project management model* seemed to have become a *discursive force* in the struggle to create hegemony in institutional practices. It enforced compliance by withholding funds. At the RDA, a team of people carried out detailed management of contracts for every HTC project. The RDA employed a 'project management model' that focused on detailed auditing of project outputs and did not reflect a comprehensive view of the issues. 'The RDA can be less constructive on the soft aspects than the hard aspects. Land and buildings are not the solution'.[39] At the micro-level of HTC projects, an unhelpful bureaucracy impeded progress because they attended the wrong metrics. They defined success in terms of *outputs* rather than *outcomes*.[40] By focusing on the things that could be measured, important factors that were difficult to quantify were ignored. For instance, there was little emphasis on qualitative changes in behaviour that could sustain meaningful change. This also highlighted a contradiction between the espoused rhetoric in the RDA's mission statement and the 'actual practices' of the RDA managers and officers. With hindsight, it seems obvious that impact of public policy and additional value of public funding in the EU could be much greater if there was less emphasis on auditing outmoded outputs and a more enlightened focus on the learning that emerges from working in new ways so that the initiatives that have been most successful can be sustained.

8.2.4 Vignette: bureaucracy – espoused and actual autonomy

In the UK, the activities of the three HTCs were governed by complex and bureaucratic approval processes, which impeded the flow of funding and restricted progress. In the implementation of ambitious innovation policies driven by public funding, the style of governance is often in conflict with the desired changes in practices.[41] In the case of the HTCs, the peculiar circumstances surrounding the Rover Crisis created an exceptionally complex array of partners and an unmanageable bureaucracy, which had a very low tolerance of risk, and this constrained the capacity for creativity, flexibility, and even obscured communications within the partnership.

Despite the policy intention in the HTCs to bridge the development gap, there was little tolerance of risk, and creativity was stifled by convoluted funding mechanisms and audit procedures. Bureaucracy hampered the innovation process. It focused on the output measures that dominate EU structural funds. It resulted in the focus being diverted from innovation metrics towards most sensitive 'political issue' associated with the Rover Crisis. Consequently, success was measured inappropriately, by the number of jobs protected or created as a result of policy intervention (Stame, 2003). Indeed, there is some evidence that the desired cultural change in the West Midlands was frustrated by one of the key policies designed to foster it. The way in which the HTC policy was implemented may have failed to bring about positive attitudes to innovation and created a suspicion of partnership and a strong aversion to risk. Not everyone understood that flexibility and complexity could not be represented as straightforward and tidy. The overwhelming impression was that people at meso- and micro-levels, and in particular project staff from universities, accepted the inevitability of working with multi-faceted complexities. This heartfelt comment from an individual working with SMEs on innovation projects illustrates the frustration this engendered.

> Reality does not correspond to a tidy plan. It is difficult to describe the flexible way we interact with businesses. Consequently the approach we adopt looks more like an amoeba than a structured business plan. The approach could be described as **'leading from below'** [emphasis in original].[42]

Serious problems arose from complex and conflicting rules in the EU funding packages, which underpinned the capital and revenue expenditure on HTC projects. For instance, the partial coverage of some parts of the corridor by geographically focused packages of ESF and ERDF support from the EU made it difficult to decide how to ensure that some geographically defined parts of the HTC were not disadvantaged.[43] The complex cocktail of ERDF, ESF, DTI, and the RDA funding made clashes between rules inevitable. The complexity of the private and public funding package had a negative impact at micro-level because 'nested rules of eligibility' make it difficult to provide timely and appropriate help to SMEs.

The HTC steering group yearned for more autonomy and financial independence from the macro-level group, who in fact made all the important decisions. The macro-level controlled the flow of funding to the lower levels and constrained their capacity for practical action. At the top level in the hierarchy, the RDA was the 'accountable agency' for public expenditure on the regional HTC policy. The 'project management model' adopted at macro-level focused on the accountability for expenditure and did not reflect a holistic view of the issues. It was experienced as unhelpful interference in operational activities and confirmed that the locus of power remained entrenched at the top level in the implementation hierarchy. 'The problem lay in the shallowness of policy formation. The RDA were deeply insecure about their understanding of the policy. They relied on outputs and rules to exert their ownership of HTCs'.[44] Although funding rules were outside the control of the regional policy makers, communication between the levels could have been smoother. There was a contradiction between the policy rhetoric of espoused autonomy for the HTC steering groups and what really happened in practice. In practice, the meso-level experienced very limited autonomy. Individual projects at the micro-level were monitored in great detail, and there was little flexibility to adjust them once they had started. If the implication of the funding rules had been communicated to the micro-level at the outset, expectations could have been modified.

A polarized 'us and them' culture could have been avoided if the complexity of the funding rules and division of labour had been clearly articulated and discussed. Even if people were not empowered to change the rules, at least they could understand what was happening and why. Opportunities for expansive learning were missed, and a culture of blame and anxiety developed. This exacerbated the separation between the micro- and macro-implementation levels. Seemingly, the boundaries between levels became stronger in order to protect the top strategic level from what they felt was a barrage of unfair criticism from the operational levels. This further reduced vertical communication between activity systems. Consequently, the capacity to make mid-course policy corrections based on feedback was reduced. There was little reflection on how policy was experienced by the 'beneficiaries'.

It was as if the *meta-activity system* at the macro-level isolated itself and sealed up communication channels to protect themselves from engaging with practical operational issues. They refused to listen to operational problems, because they simply did not know how to solve things, because the RTF funding rules allowed very little flexibility. This analysis was strongly endorsed by key actors who had been involved in the HTCs from their inception.

> The RDA were insecure and lacked confidence in their role in HTCs. The Universities had taken the view that the RDA was a fragile beast at the time and that it would be unhelpful to be too critical as HTCs were developing. In 2000 the Universities in the region decided to nurture the fledging RDA.[45]

Instead of openly sharing the problems and having transparent discussions with the operational teams, progress stalled, and those at the macro level, who understood the reasons for this, remained silent. The failure in communication fractured relationships of trust. This wasted time and dissipated enthusiasm, and with hindsight, sustainable collaboration occurs on the basis of openness and shared understanding.

By contrast, the policy rhetoric raised expectations about 'working in partnership' and increased agency at the meso-level by passing the responsibility for financial decision-making down the implementation chain to HTC steering groups.[46] This caused frustration because the devolution of autonomy for decision-making to the subregional steering groups was slow and partial.

> Nothing much has changed as yet. For instance it took two and a half years to negotiate the NOVUS project. It was a cocktail of funding packages that made it difficult to operate out of the Objective 2 areas of the CTB.[47]

The practice of *working in partnership* was at odds with the *rhetoric of partnership* espoused in regional policy documents, in publicity brochures, and in the public Web sites, which described the HTCs as autonomous partnerships. In this instance, acquiescence to the dominant partner was misrepresented in public as *consensual praxis*, and the foundations of expansive collaboration were undermined.

In other words, the macro-level constrained the emergence of new practices at the micro-level, which were proposed because of practical experience with local 'beneficiaries' of the policy. Consequently, the object of activity at the macro-level was in tension with the local operational practice based on detailed knowledge of the context and circumstances in which SMEs worked. This disempowered people working at the operational (meso and micro) levels from implementing the HTC policy in the way that fitted the local context best.

There was a contradiction between the strategic plans produced at the macro-level in the policy implementation hierarchy and the need for operational flexibility in the HTC projects. It was difficult for local discourses and feedback from SMEs to percolate upwards.

There was no evidence that boundary objects had been created in the HTCs to bridge the vertical gaps between the perspectives (and objects) of decision-makers at macro-, meso-, and micro-levels. It is difficult to depict and explain the problems with vertical communication in terms of expansive learning theory, because the concept of 'organizational collaboration' is not well suited to depicting 'a chain of command'.[48]

8.2.5 Vignette: the creative power of conflict – action on sustainable energy

This next vignette from a case study in the USA concludes this chapter on a more positive note by describing some practices that reduced the impact of the dominant partners. SMUD was driven by the profit motives of shareholders but over the past thirty years public interest research and community involvement has transformed energy policy in Sacramento. Instability arising from contradictions between different perspectives can be seen as a 'creative' stage in the process of change. Responses to these tensions can drive change in a particular direction. By raising awareness of the process of 'object construction' and the trajectory of activity, it is possible to see conflict differently. 'Conflict engenders new and better ideas and pushes towards a more comprehensive understanding' (Hellström, 2004: p. 639–640). The following vignette illustrates the positive effect of conflict and debate in driving innovation. However, it is also clear from this instance that listening to minority view from a relatively weak activity systems was very important to sustaining the trust and the long-term engagement of partners. The chances of innovation seem to improve when people with opposing perspectives meet in a constructive climate

> – 35 years ago SMUD built an early nuclear plant – it was one of the less successful designs and had problems. Conflicting interests in the 1970's and 80's made some sparks fly and changed public opinion about nuclear technology. Due to technical problems with the plant, SMUD at the time tried to refurbish it – but due to a strong anti nuclear movement & lobby, the SMUD Board was asked to place on the ballot whether the plant should be closed. The public voted to close the plant at a time when the repairs were just completed, and more than. 50% of the generating capacity of SMUD was lost overnight. SMUD's assets value and bond rating dropped precipitously, the cost of borrowing rose significantly, and millions of dollars in repairs done to the plant were lost.
>
> (M. DeAngelis, interview transcript, 12 October 2006)

In this example, several worlds collided in a constructive way and conflict had a positive impact on moving people out of their 'entrenched position' towards a consensus. It took a long time, but it was a great way to resolve a highly emotional and politicized problem.

8.3 Overview

The vignettes showed that organizational learning is constrained by complex bureaucracies. They show how the HTC approach to innovation stands in contrast to the examples of innovation at Google and Nuno described in the vignettes in Chapter 7. In the HTCs, the complex rules and regulations associated with public funding and long-term politically sensitive goals made people risk averse. It created a climate where rigid structures and processes caused acute anxiety about failure and impaired innovation capacity.

There is no magic formula to ensure successful innovation, but evaluation can be a powerful tool for learning. Policies for innovation and technology transfer are emotive and politicized, and when people work across boundaries, they enter *hybrid* and potentially hostile territory. Claims for ownership of the 'territory' of technology transfer are many and various. The vignettes showed how activities draw people into 'contested terrain' because individuals and organizations are working in new domains, without the comfort of familiar routines and practices.

Vignettes from the HTCs explored the relationships between apparently contradictory phenomena that cause tensions between activity systems and drive change. These contradictions highlighted differences between 'how the implementation of the innovation policy was experienced' and 'how the process was intended to be'. Despite the frustrations and problems, the data indicated that people at the micro- and meso-levels worked together to overcome the bureaucratic blockages to progress. Overall, respondents were reluctant to be critical, and there was consensus that everyone was doing his or her best in uniquely challenging circumstances. Although this did not alleviate a sense of frustration, it did indicate a commitment to make HTC policy work in the UK despite the past and present problems. I will turn now to Chapter 9, which presents an overview and summary of my research findings.

9 Insights from the research findings

Four research questions encapsulated the aims of this book.

Research Question 1

- Why is it that some approaches seem to have evolved in a particular geographic and social context?

The vignettes focused on specific differences in the scale and structures of collaborations, which brought industry, government, and universities closer together to implement technology transfer. Both Japan and the UK have attempted to catalyse their own 'Silicon Valley' success stories through public funding and policy interventions. However, various incompatible fragments from diverse theories of change seem to have shaped policy (Weiss, 1995).

The case studies varied in their complexity. The simplest examples described innovation projects driven by a single company; the most complex example of innovation described a raft of programmes and projects launched by a very ambitious innovation policy in the UK.

In the UK, the complicated context of the implementation of high-technology corridor (HTC) policy was due to a confluence of three circumstances. First, the crisis for automotive manufacturing in the region focused attention on job losses resulting from the threatened closure of the Rover plant. Second, the intervention of the Department of Trade and Industry (DTI) and the establishment of the Rover Task Force (RTF) aligned the regional public and private sector stakeholders to address the ensuing problems. Third, the opportunity arose to add value to DTI and Regional Development Agency (RDA) funds by matching it with European Regional Development Fund (ERDF) and European Social Fund (ESF), which created a funding package of Byzantine complexity with contradictory rules and foci.

The HTC policy assumed that copying models, that had worked successfully elsewhere, could create a regional knowledge economy. This international practice of 'policy borrowing' was discussed in Chapter 2. Policy makers assumed that 'technology transfer' could be achieved by adopting the ideas and mechanisms that had been effective in Silicon Valley, Massachusetts, Stockholm, and Cambridge.

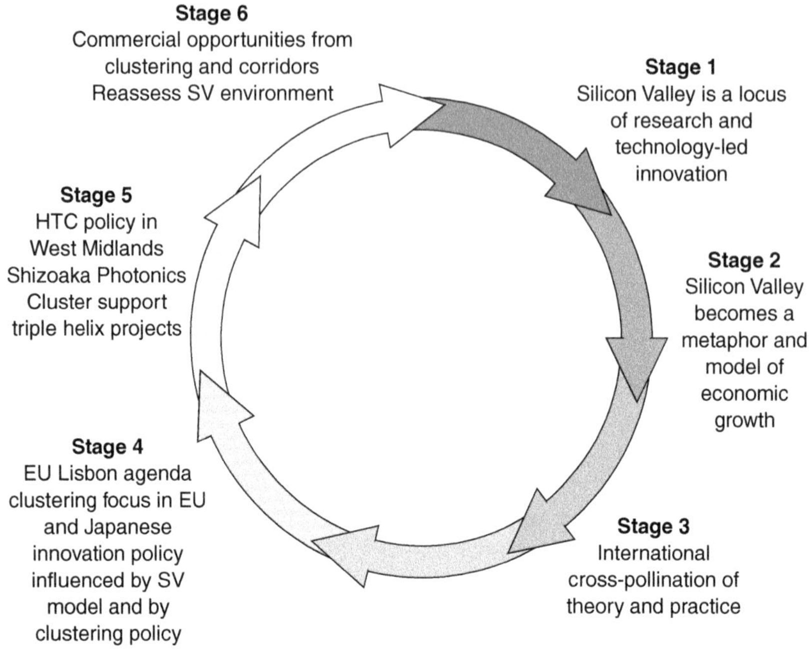

Figure 9.1 Policy borrowing stimulated by the Silicon Valley phenomenon.

Implicitly, this underestimated the human dynamics and communication gaps that influenced whether 'technology transfer' was successful or not.[1] The mutual influence of innovation policies stimulated by the Silicon Valley phenomenon can be depicted in Figure 9.1.

The creation of the photonics cluster in Hamamatsu followed a similar 'policy borrowing' rationale. It was stimulated by a government policy backed up by public funds as an incentive to catalyse 'triple helix' engagement. The implementation of the regional clustering policy resulted in large-scale innovation projects that could not have been achieved by any of the partners working in isolation.

Although the successful growth of a cluster of high-technology companies in Silicon Valley has been used as a template for innovation elsewhere, vignettes from my case studies focused on the 'environment' in which that success took place. In this part of the USA, it is clear that unique financial conditions and service industries have grown as a response to university-led innovation in Silicon Valley. This has created a sustaining environment for start-ups and high-growth companies, which is very difficult to replicate in less market-driven economies. In Silicon Valley, the flow of knowledge was driven by market forces rather than by government intervention.

The second US case study showed another facet of the circumstance that inhibits or fosters innovation. It took a major energy crisis to unlock the potential of alternative sources of energy that had largely been ignored due to vested interests and general complacency. Figure 9.2 shows how the recent changes in

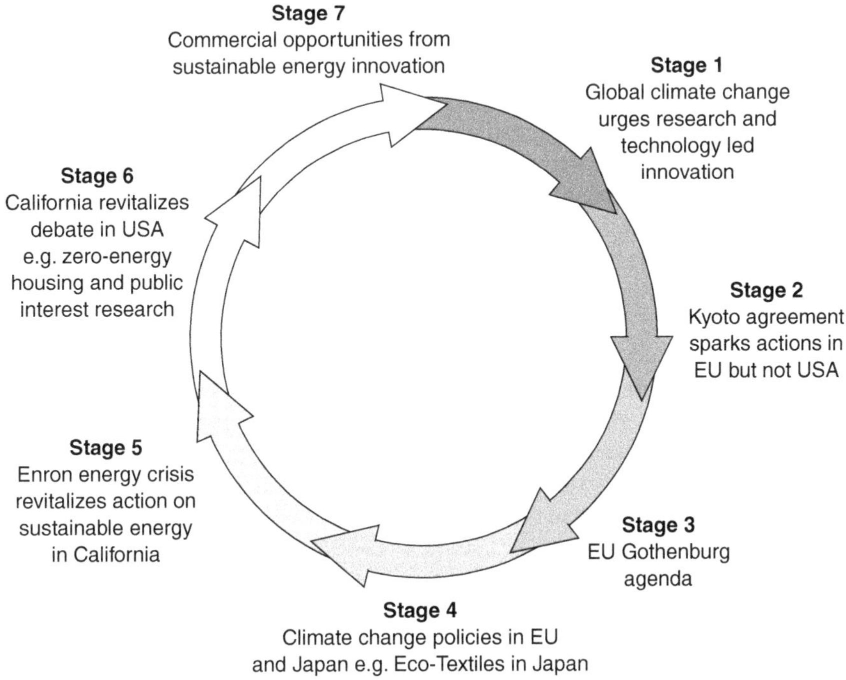

Figure 9.2 Cross-cultural influences on energy policy.

environmental policy has resulted from some cross-pollination of idea tries and tested in different counties.

The 'community-oriented' approach of the Sacramento Municipal Utility District (SMUD) offers a hopeful indicator of the positive effects of a growing public interest in environmental issues. The region around San Francisco in California is not typical of the USA in general, but nevertheless, it can catalyse change by example.

Research Question 2

- What can be learnt from such exemplars and what if anything can be transplanted or adopted in a different context?

The Economic and Social Research Council (ESRC) and other funding bodies devote serious consideration to the meaning of 'relevance in practice' (Rappert, 1999). It is important to appreciate that the perspectives of policy makers and beneficiaries such as small and medium enterprises (SMEs) tend to be incompatible. These two groups may be interested in different insights from my findings and may interpret them differently. The stimulation of innovation through government-led policy initiatives has been shown to involve convoluted multi-agency structures. The problems associated with a lengthy implementation chain were highlighted in vignettes from the case studies. In particular, this book may help policy makers 'to understand the perspectives of people implementing policy in practice'.

Rapid and large-scale organizational change is extremely challenging. Opportunities for reflection and learning can be severely limited by highly politicized and risky endeavours. History and previous cultural practices are critically significant factors in the implementation of an innovation policy. A huge implementation gap can arise in complex policy, and the quality of the human dynamics is crucial to the success of the implementation process. An enormous amount of time and effort is required to build partnerships. People are frequently disconcerted to find that the object, on which they thought they had agreed, turns out to have transformed itself into something different.

Pilot projects are a valuable tool for gaining experience and prepare people for implementing something more complex or large scale. Yet, in the UK, there was an urgent need to do something to take advantage of a unique funding opportunity for developing a regional knowledge economy. Inevitably, this meant starting work before setting a clear and carefully considered strategy. Unsurprisingly, there were instances of people working valiantly but not necessarily in the same direction. In expansive learning theory, this would be expressed as being orientated towards different *competing objects of activity*. In everyday language, this might be expressed as misalignments of stakeholders, which left individuals confused as to how to work together.

The vignettes illustrated that a lack of opportunity to develop experience created organizational 'blind spots'. Pilot project can help to test out the actual effect of a policy before a full-scale launch and reduce its vulnerability to failure. The rapidity of a response to the Rover Crisis did not allow any time to learn from pilot projects, which resulted in a full-scale launch of an untested programme.

The espoused policy portrayed in publicity materials such as the West Midlands Regional Innovation Strategy and Regional Economic Strategy documents assumed that the key was to put groups of 'developers' and 'users' together. Initially, the policy makers in the West Midlands did not prioritize the need to actively help people to work together. It illustrates that 'change' cannot be 'taken for granted' if organizations have no prior history of working together. The difficulty of unfreezing entrenched positions and moving towards a shared vision were underestimated (Trowler *et al.*, 2002).[2]

Surprisingly, the greatest polarization was not between public and private sector organizations. The strongest differences in 'technology transfer' practices depended on how close the individual was to the practical application of a new idea or technology in a local company. Individuals who could draw on practical experience of technology transfer made the most insightful comments. They described the process as fraught with unpredictable complications and occasional serendipity from which unexpected benefits might flow. They stressed that it was important to nurture long-term relationships and to invest time and effort as well as resource.

In the UK, policy makers assumed that government interventions were capable of stimulating the same kind of clustering of high-growth companies that had flourished in Silicon Valley and Boston. However, there is evidence that the bureaucracy associated with public funding can choke the entrepreneurial spirit it aims to foster.

These empirical data challenge current theories and tacit assumptions about how complex business and university multi-agency partnerships 'work in practice'. It questions the assumption that if government acts to bring together business 'users' and research 'developers', the desired change will be achieved. Expectations inspired by policy rhetoric in the UK were far from what actually happened in practice. Contradictory behaviours and mixed messages caused confusion and created a culture of anxiety and blame. In general, there was inadequate debate, and in some instances, *consensus* was forced.

In the absence of mediation, this can result in 'overt compliance' arising from an enforced consensus and the suppression of alternative views and tends to distort potentially collaborative relationships (Fullan, 1993: p. 82). In the absence of negotiation between the perspectives, those people in the most powerful positions simply issued instructions to the meso- and micro-levels. 'The RDA felt threatened and became increasingly defensive and critical of the people at the delivery level. In turn those people working at the delivery level became increasingly frustrated'.[3] Acquiesce is not the same as emotional commitment and does not lead to sustained cooperation. Chapter 7 described how partners drew closer together to fight 'coercion'. The data showed that an HTC steering group worked around the problem of 'powerlessness' by setting up a company limited by guarantee, which gave them more autonomy.

By alerting people to the tensions that would otherwise be ignored or tolerated, a process of dialogue, experimentation, and collective learning can be triggered that may transform participants' understanding of their activities and the systems through which they are enacted (Blackler, 1995).

Organizations responsible for using public funding to catalyse economic development may benefit from a more reflective stance and systematic evaluation of the effectiveness of policy. This would help policy makers understand how policy is 'experienced' in the implementation process, without having to be defensive about public money.

Research Question 3

- What are the benefits and limitations of using expansive learning theory to analyse the development of multi-agency learning?

An analysis of the tensions in complex activity networks offers a *stronger theoretical perspective* than mere descriptions of controversies or divergent views. Expansive learning theory helped to identify dynamic conflicts in the relationships between people and agencies and the concurrent impact of social structures including institutions, rules, and norms. In order to be innovative in a larger social system, the innovating group or person has to be 'incubated' in another social system. This means that role expectations, and norm-driven actions, would be innovative in comparison with the outside social system. This is relevant to the contradictory conception of the innovating entrepreneur as being both a generalized cultural type and a pure expression of individual intentions (Hellström, 2004: p. 640).

This research illustrates the importance of *figure and ground* in that individual human relationships are *mediated* by tools, communities, rules, and divisions of labour orientated towards an object of activity. In turn, relationships within and between activity systems are also *mediated* by proximity to networks or *meta-activity systems*.

Human relationships can be described theoretically in terms of *meta-activity systems* or networks in which sub-activity systems collide and interact vertically as well as horizontally. All these relationships are mediated by tools, social rules, and divisions of labour orientated towards an object of activity and collaborative outcomes such as organizational learning.

Analysis suggested that mechanisms to link the activity systems at various levels could help the implementation of innovation policy to be more effective. In the terminology of expansive learning theory, the term 'object' encapsulates issues concerning individual and shared interests; the divisions of labour and rules are associated with organizational positioning and 'perspective'. The horizontal connectivity tends to occur between experts, but the vertical connectivity between policy makers and the people taking action at the grass-roots level is much more difficult to achieve. As yet, it is not well described in current versions of expansive learning theory, and it is this gap in that I set out to explore.

Recognizing the existence of competing objects of activity can stimulate healthy debate and a consequent realignment and expansion of perspectives. The research suggests that it is helpful if there are realistic expectations about conflicts and tensions. In particular, the vignettes have drawn attention to the difficulties that arise when contradictory sets of rules exist in different parts of the implementation chain.

In expansive learning theory, collaboration and multi-agency learning depends upon individuals sharing their ideas and working towards shared goals, but this has led to criticism of a perceived tendency to reify organizations (Thompson, 2004: p. 583). In other words, critics argue that the tension between individual and collective consciousness is not completely resolved in theoretical terms. It is therefore suggested that the theory has been stretched too far and extrapolating activity theory from its origins in individual consciousness and learning, which may weaken its conceptual coherence.

At the root of the dilemma of clarifying the 'subject' in activity is the drift of organizational activity theory, away from 'its moorings in Vygotsky's original focus on human consciousness' (Thompson, 2004: p. 580). When organizations become reified at group level, this contradicts the original premise of activity theory that organizations emerge through interpretative human activity. By reifying organizations and groups, they acquire an ontological existence. It is therefore suggested that this converts a dynamic process of organizational development into a representation, which reverts to the Cartesian dualism, between subject and object, that activity theory originally sought to overcome. Expansive learning theory accepts that the self changes constantly. Historical selves are entities who 'create futures yet are themselves created by the past' (Hellström, 2004: p. 643). In the attention to boundary crossing, the theory encompasses the melding of multiple identities and transient social worlds in the process of innovation.

In my case studies, it was very important to clarify the meaning of the term *subject*. The scale of interactions of subjects from different activity systems varied considerably. At one end of this spectrum, vignettes described collaboration between two companies on a discrete innovation project. At the other extreme, vignettes described complex innovation programmes and interactions between numerous complicated activity networks. The findings suggest that *self-identity* is created by continuous interactions between an individual and social worlds. Individuals acted in one or more activity systems across organizational cultures, but they were also members of a tier in the policy enactment hierarchy. Membership of horizontal and vertical activity systems was obscured because it changed over time. Although these activity networks comprised complex multi-layered, organizational partnerships, it was possible to describe individual and collective 'subject perspectives'. This was an interesting phenomenon, yet it did not necessarily reify any group or organization. (Reification was discussed earlier in Chapter 2.)

Despite the dilemma of using graphic depictions of an essentially dynamic process, I used diagrams to illustrate links between the data and the theoretical analysis throughout the text. Also because expansive learning theory is evolving, there are few precedents for studying organizational complexity. I therefore have had to create my own diagrams and to explain what the data seemed to be suggesting.

Expansive learning theory frames the contexts and circumstances of inter-agency learning and draws attention to the circumstances in which new collaborative practices are created. Although expansive learning theory may not be able to predict the best course of action, it can draw attention to factors that are likely to affect the successful implementation of similar policies elsewhere.

Research Question 4

- How can boundary objects and mediators facilitate innovation?

Special tools called *boundary objects* helped to connect experts horizontally and were particularly useful if there was no previous experience of working together. They acted as bridges between different cultural practices. They enable people operating at the same level to learn from one another by expanding the 'object of their activity' with a common vision. There was evidence in my data that the creation of a *boundary object* helped to 'crystallize emergent practice'. The creation of a boundary object helped individuals from different organizational cultures to share a common vision and a sense of identity as a new 'community of practice' with a sense of shared identity and allegiance.[4] Boundary objects captured 'transitory insights' about how the partners could work more closely together. At the same time, the *use* of the boundary object buttressed the emerging collaborative partnership. For instance, when the Web site boundary object was used by the HTC steering group, it strengthened the confidence and trust within the partnership as described in vignette 'collaborative projects to develop electronic tools'.

In Chapters 5 and 6, I described some boundary objects that catalysed closer working relationships between users and developers of specific technologies. Lateral boundaries between activity systems restricted movement across cultures, but they were comparatively weak. Sideway moves *across* boundaries were evident in many instances. For instance, it was difficult to get clear and coherent messages about 'technology needs' from local companies, so mechanisms to smooth communication between *user* and *developer* communities could be very constructive. Nevertheless, my data suggest that it is not possible for users to specify their demand if they are unacquainted with what is available.

For example, collaborative visualization software and rapid prototyping equipment acted as boundary objects. Several respondents described boundary objects as tools that seemed to have 'crystallized collaboration' in the sense of structuring routine ways of working together.[5] The phrase *we now have a tool to do x* appeared several times in the data. This supports the view that policy enactment is smoother where the process is supported by 'quality materials' and 'a highly interactive infrastructure of pressure and support' (Fullan, 2001). Working across boundaries and sharing the creation of boundary objects involved sharing common issues and problems. Boundary objects have agency and are important tools for innovators because they can move people towards new ways of doing something. For instance, the innovative applications of a new technology can more easily be understood by reference to an artefact.

Boundary objects are versatile and can act as tools to facilitate communication between separate activity systems, but they can be transformed by being used in new contexts by a larger activity network. As described in Chapter 6, collaborative visualization software was used as a boundary object to connect communities of experts in a large engineering company. It was then adopted by the Advanced Engineering (AE) HTC project. It then acted as a device to enrol small engineering companies to join the AE micro-cluster. When the small companies used the software, it acted as a *tool* to help them carry out new product development activities. Finally, the software was an *outcome* of the expansive activities when it was launched as a commercial product as described in 5.3.5 vignette: 'how users transformed an outcome into a tool'.

During the data analysis cycles, I began to see patterns to suggest that certain HTC projects incorporate elements of human interaction. In such instances, a boundary mediator might use a boundary object to communicate 'technology transfer' opportunities to SMEs. I had not previously considered the potential interactions of boundary mediators with boundary objects.

However, I found that special people at the macro-level of policy enactment level can act as *boundary mediators*, and in this role, they can use a *boundary object*. At first, I thought this was a curious and contradictory finding, because my first analysis of the data suggested that boundary objects linked activity systems horizontally, whereas connecting levels was more difficult. It seemed that to link activity systems vertically required a boundary mediator. I was able to check this with key actors who agreed that boundary mediators played an important role and recognized that in some ways he or she had acted in this role from 2000 to 2005.[6]

On more detailed analysis of the data, I found that boundary mediators had a crucial role in projects as 'technology translators'. They operated in a *hybrid* environment between the developers of a new technology and a small company to apply the innovation in a commercial process. In other words, they were not empowered as strategic actors, and therefore, they could not carry messages from the operational level, in the way that the most powerful actors could (as described in Chapters 6 and 7).

A boundary object seems to have variable agency when used in different contexts. When used by those who shared in its creation, it seemed to be more powerful than when it was used by other activity systems where it had limited and general value as a communication tool. The emotional investment and sense of ownership that happens in the creation of a boundary object optimizes its functionality as a tool. This suggested that mediation by boundary objects and boundary mediators was not a black-and-white issue but more a question of *degree*. This suggests that a boundary object can have different degrees of *agency* depending on the user and the context of use. Actors who had shared in the creation of the boundary object experienced it as a powerful tool. Although other actors could use the boundary object, it was not such a powerful tool for them because they had not developed it. The data showed that the creation and use of Web sites and electronic databases helped to build upon shared knowledge and experience and new divisions of labour, and this concurs with findings in a recent study of innovation in Scandinavia (Hasu, 2001: p. 49). However, I found no evidence of a boundary object being created simply to link two or more levels. It seemed that motivation to create a boundary object between levels was weak. 'The people responsible for strategy made little attempt to understand what was happening at a practical level in the HTCs'.[7] Power differences between activity systems were dysfunctional to a commitment to harmonious collaboration. Those at the macro-level did not need to invest time and effort communicating with those below enacting the policy at a practical level. The character of relationships between levels was hierarchical and instructions tended to emanate from the macro-level with little opportunity for negotiation.

Projects can be relatively small and involve a few partners. On the other hand, projects can be large and complex and involve a vast range of organizations. In some instances, small projects may also act as tools for a larger project. Complex projects can use smaller projects as 'embedded boundary objects' rather like the way expansive learning builds on small expansive cycles. In this way, projects draw on past projects and expansive cycles, and *boundary mediators* can use such projects as *boundary objects* to make their task easier.

Innovation projects seemed to be most effective when they combined the work of boundary mediators with one or more boundary objects.[8] Vignettes 'collaboration in the creation of eco-textiles', 'uniting communities in collaborative public interest research', and 'embedded projects within Medilink' illustrate this particularly well.[9] Two boundary mediators who had strategic influence at the macro-level of HTC policy were successful in lobbying for project resources. The Medilink collaboration was built on the success of previous, small collaborative projects. For instance, the Medici project acted as an embedded boundary object for more expansive collaboration in Medilink. The Medici was a collaborative project for universities to train medical and bioscience researchers in entrepreneurship to help them to turn research ideas into commercial opportunities. Medici commissioned research into the potential for a medical technology cluster in the West Midlands. This research also acted as a boundary object to connect the steering group and unite them in allocating resources to Medilink projects (Burfitt *et al.*, 2003).

While respecting the cultural practices of each organization, Medilink mediators can help people to orientate their efforts towards expanded objects of activity. Boundary mediators have a privileged position and can communicate the problems and issues from different organizational perspectives. They have the power to translate and communicate messages from the micro-level to the macro-level that would otherwise be ignored or blocked.

With respect to small-scale innovation projects, vignettes 'industry scanning and serious play – jellyfish fabric', 'metallic plating of textiles', 'the capacity of designers to act as mini-maestros – holograms', and 'collaboration in the creation of eco-textiles' describe the work of the Nuno Corporation (Sudo, 2006). These examples support the notion that boundary mediators play an essential role in translating the practices from one industry sector to another. Mediators can use 'samples and prototypes' as highly effective boundary objects.[10] An understanding of different industries enables mediators to span industrial sectors and apply a new technological process or product developed in one industry in a new discipline and commercial context.

Research Question 5

- What insights from expansive learning theory and from the analysis of collaboration in these international case studies can be applied to similar policy initiatives?[11]

It cannot be assumed that 'partnership' occurs as a natural and inevitable consequence of bringing together people from the private and public sector with resources to achieve a complex policy goal. The data show that without structures to support the robust exchange of views, *collisions between perspectives* fail to expand the object towards collaborative activity.

Partnership has been shown to be a much-abused concept in policy. At the root of the difficulties of implementing a complex policy was a mismatch between espoused autonomy and real agency. The vignettes from the UK case study showed that a disassociation of power and accountability was unhelpful. Ironically, the power dynamics between implementation levels makes it difficult for the voices of the 'disempowered' to be heard. This tends to omit the perspective of people closest to 'technology transfer' practice and inhibits feedback about implementation. Policy makers placed too much emphasis on the financial responsibilities of steering groups and insufficient focus on creating time and space for high-quality networking. Insufficient time and effort was invested in communication and the shared development of new practices.

Expansive learning is impaired when there is strong pressure to achieve 'success' very quickly. Collaboration is difficult when success is defined by 'target-driven' performance measures linked to tangible *outputs* rather than sustainable *outcomes*. There is a danger that when linear progress is not evident, a promising experiment is curtailed and a new raft of policy is imposed before the original policy has been given time to bear fruit. This tends to be counter productive and arises as a 'quick fix reaction' to fear of failure.

My research shows that when organizational cultures create an atmosphere of anxiety, individuals tend to protect themselves by avoiding blame. Partnerships risk spreading responsibility so thinly that nobody is accountable for anything specific. Conversely, it also seems to be dysfunctional to collaboration to project, manage, and audit every minute detail of activity.

The collisions between activity systems operating at different levels did not result in multi-agency expansive learning because of inherent differences in political power and agency. Expansive learning between activity systems occurs when they learn from one another through mutual collaboration. My findings suggest that the dynamics of power can impair collaboration. The collisions between the activity system of the policy intention or theory (macro-object) and the activity systems of the practitioners (micro-objects) did not lead to expansive learning. The hierarchical implementation structure frustrated collaboration and mutual learning. The theory does not yet deal adequately with the issue of motivation and power.

> In terms of power and motivation the Universities saw HTCs at the time as having three possible benefits. 1) HTCs cemented their regional role 2) it played into their national role defined by HEIF 3) there were some opportunities for capital equipment costs to be met through HTCs. Universities had a unique role beyond the horizon of RDAs and were themselves a driving force for investment.[12]

This suggests that it is vital to consider *intention* in the study of innovation, including mediation of identity, technological factors, emotional commitment, and re-conceptualization of the object of activity. As depicted in expansive learning theory, the creation of sustainable collaborative practices seems to be exceptionally difficult (Tuomi-Grohn, 2003: p. 207).[13] Small expansive cycles seemed to be most effective in changing practices throughout the HTCs. The data showed that although a project was a 'small step' or a miniature expansive cycle, it can act as a catalyst for more ambitious and extensive work. Currently, the theoretical vocabulary is not sufficiently developed to describe the variation in the quality of collaboration and the commitment.

The term of 'developmental transfer' is now commonly used to describe collaboration in work contexts. Although developmental transfer emphasizes the impact of history, rules of engagement, and divisions of labour on the ways in which organizations collaborate, it does not address the variation in the quality of collaboration. For instance, a number of respondents described the HTC programme as 'a partnership'. However, the term meant various things to different individuals.

The word 'partnership' appears frequently in the policy discourse, but the quality of engagement of the partners is likely to be seen in terms of 'action' and 'practices' rather than rhetoric. It is important to understand the dynamics of 'partnerships' and to differentiate the quality of commitment to the 'object'. Activity theorists often use the term 'advanced collaborative practices' to describe sustained and progressive interaction. However, collaboration has two alternative meanings. In some contexts, it implies a treacherous alliance with an enemy, but in other contexts, it is used interchangeably with the term cooperation.

In everyday usage, the term collaboration suggests strategic alliances towards goals of mutual interest. Cooperation captures the essence of expansive learning and *partnership*, associated with a commitment to shared values and a vision that motivates and sustains shared work (as discussed in Chapter 7). Cooperation happens when projects are anchored in wider practice. Distinctions can be made in the quality of relationships and suggest that:

Cooperation describes the sub-regional focus on local problems adopted by the members of an HTC steering group operating at the meso-level and interactions in HTC projects where co-configuration occurred at the micro-level.

Instrumental collaboration describes the macro-level 'strategic alignment' of regional agencies to solve the 'Rover Crisis'.

Pragmatic acquiescence describes the reluctant agreement to something that seems to be inevitable. This was manifested in the acceptance of delays in funding and complicated auditing procedures that resulted in tension between the RDA (as the 'gatekeeper and auditor' of funds) and the other partners.

Less rhetoric espousing partnership and more practical demonstrations of collaborative decision-making may have worked more effectively. In one of the rare references to the dynamics of power in expansive learning theory, Engeström explains that the espoused object of activity is creatively reconstructed. This issue is at the heart of the clashes between the macro-, meso-, and micro-levels of HTC policy enactment.

> No matter how clear the intention and assignment may be for management, **the object** will be creatively reconstructed by those who are supposed to

> solve the problem. This creative reconstruction often involves questioning, confrontation and debate. If this is overlooked, the important dimension of **power** will be artificially separated from object-oriented collaborative work and innovative learning in work organizations and teams [emphasis added].
>
> (Engeström, 1999: p. 392)

Expectations about collaboration outstripped what was actually possible, and it was unrealistic to represent the notion of partnership as an easy process. At the meso- and micro-levels, there was a lethal combination of having responsibility for outcomes of policy while at the same time being powerless to change the way things were done. The HTC directors were tasked to write strategic plans. It was expected that these local plans would predict the future development of the HTCs. In this sense, strategic plans are close to 'theories about what needs to happen next' (Colville *et al.*, 1999). However, the environment for HTCs was so unstable that strategic planning was fraught with unpredictable complications. 'The demand for outline budgets at short notice did not recognize the time it takes for a partnership to go through the iterations of a development plan'.[14] Partnerships are not harmonious and naturally perspectives collide. Disagreements can be creative and are the drivers for expansion and collaboration. Suppression of dissent is damaging and can create Shibboleths of partnership. Acquiescence is a poor substitute for passionate commitment. 'Innovative learning processes may not be harmonious brainstorming sessions or situations where "members think alike"' (Engeström, 1999: p. 400). Processes of innovation and knowledge creation exhibit expansive and non-expansive phases; both step forward and digress (Engeström, 1999: p. 391). Discussion and disagreements can result in collective refocusing, even though the object is captured in *fuzzy* and *ambiguous* form (Engeström, 1999: p. 392). If a complex programme can treat embedded projects as experiments, it is possible to learn from them and build trust and mutual confidence to enable more ambitious collaborative work to take place. This finding relates to Engeström's emphasis on the *Collective Zone of Proximal Development*, and it suggests that wholesale ambitious change, under pressure of time and money, is unlikely to be successful in one step because the gap between old and new practices is too great.

10 A toolkit to support organizational learning

10.1 Practical ways to enhance innovation partnerships

As I have argued throughout this book, partnerships and collaborative working practices form a bedrock idea for change and growth in organizations. This chapter attempts to identify some tips that might support change within this kind of paradigm. The dominant policy discourse portrays the notion of 'partnership' as a panacea for the solution of entrenched social and economic problems. However, overuse of the word 'partnership' has clouded its meaning as discussed in detail in Chapter 9. Policy makers tend to use the terms collaboration and partnership as a general catch-all. Economic and social development policies assume that 'partnership' simply happens if people from different organizations are persuaded or coerced into working together to deliver a desired change. This assumption ignores the risks and complexities inherent in change, which is characterized by insecurity and anxiety.

In the previous chapter, I focused on important differences in the style of multi-agency partnerships. I suggested that the most developed and sustainable type of partnership is characterized by 'cooperation'. By contrast, the most superficial and insecure partnerships are characterized by 'pragmatic acquiescence' in which weaker partners give way to the dominant actors. However, many partnerships fall between these two extremes. These can be described as 'instrumental collaborations' in which 'a loose alignment of organizations' works towards a strategic goal that could not be attained independently.

It is important to recognize that partnership is a process. Partnerships evolve and change over time. In other words, a description of a partnership may be valid at the time, but it is unlikely to stay that way forever. A description of this type is similar to a snapshot that records the practices adopted by a partnership at a particular moment. By contrast, a longitudinal study might capture the evolution of a partnership over several months and may be considered to be like a short video.

It is helpful to ground this discussion in practical examples that illustrate the developmental history of a partnership. Vignette 'how users transformed an outcome into a tool' described how 'instrumental collaboration' evolved into sustained cooperation. It described how OPM software was created as an in-house tool to integrate the user perspective into new product development. However, collaboration with their 'supply chain' of small engineering companies enabled them to undertake beta-testing of OPM prior to launching the software as a commercial product.

Conversely, it is equally possible for a partnership to backslide into pragmatic acquiescence characterized by 'organizational bullying' by the dominant partners. Vignette 'bureaucracy – espoused and actual autonomy' illustrated how problems in the high-technology corridor (HTC) partnerships arose from differences in power exacerbated by long and complex implementation chains. In other words, a partnership is never completely stable, and the effort required to sustain the process should not be underestimated. It is more realistic to think of partnership as work in progress than a finished product.

Increasing the number of people involved in creating a new practice is complicated and communication difficulties arise. Where people feel empowered to create solutions to problems, their ownership of the change reduces the risk of new practices being rejected (Trowler *et al.*, 2002). Unfortunately, many recent changes in work practices have been driven by economic pressures to do 'more with less', and so many people resist change on the assumption that new ways of doing things will inevitably be worse for them. Pressures from increased performance and target-driven work cultures can make it difficult for new communities of practice to form.

Partnerships between organizations where the cultural perspectives are very divergent are consequently more challenging than those where their organizational cultures are similar or where there is a past history of working together. In creative industries and engineering, 'boundary objects' are extremely effective tools for collaboration. They can be used in different ways to align perspectives and can be used by facilitators to move a 'stalled partnership' into a more productive stage of development.

Change is more sustainable and ultimately effective if not taken in radical leaps. It also suggests that organizations can be thought to mature, just as individuals mature, so that readiness for change is important. Expansive learning theory (ELT) offers some analytical tools to understand the challenge of partnerships attempting to implement organizational change. As discussed in Chapter 2, ambitious change can be too challenging. If the gap between old and new practices is too wide, the change will falter. In a Collective Zone of Proximal Development (CZPD), the difference between old and new ways of doing things is 'just right', so activity systems can expand their horizons towards a shared object of activity.

The next section will draw on practical illustrations of partnership from vignettes that are featured in previous chapters to focus on the ways in which facilitators can act to steer partnerships through difficult stages in the development.

10.2 Diagnostic tools for innovation partnerships

The health and sustainability of a complex partnership cannot be taken for granted. This chapter proposes a heuristic toolkit in the form of diagnostic questions for multi-agency partnerships. These heuristics or rules of thumb can be used as a quick health check on the partnership arrangements and innovation potential. They do not prescribe solutions, but they can be used to diagnose potential weaknesses and areas where it would be wise to invest time and effort.

ELT suggests that *conscription devices* can help to align disparate organizations towards sustained cooperation. In other words, emotional commitment to concerted action can work as a conscription device, which aligns and sustains the engagement of the partners. For instance, the case studies showed that people cared deeply about health and environmental issues. Vignette 'uniting communities in collaborative public interest research' describes how disparate communities came together in Sacramento to fund public interest research in sustainable energy. The HTC case studies illustrate how an emotional commitment to improving health care worked as a conscription device for partners, working on innovative medical technologies as described in vignettes 'collaboration between physicians and automotive designers' and 'collaboration between defence industry and physicians'.

The Japanese case studies illustrated the impact of a powerful 'conscription device' in textile innovation. In the Nuno Corporation, the emotional commitment to 'creativity' was very powerful. It should not be undervalued as a conscription device that develops and sustains innovation partnerships. The inspirational catalogues of work produced by the Nuno Corporation illustrate this perfectly. During the past decade, Head Designer and Chief Executive, Reiko Sudo, created a series of beautifully illustrated books describing the creation of textiles around six themes.[1] These books express her artistic vision and also describe the technical innovations underpinning the creation of new fabrics and their connection to the history of Japanese textiles. Each has a name and a story that expresses her creative intention and the emotional response the fabric aims to evoke. She approached the task in a unique way and collaborated with poets, architects, and archaeologists. Most of these textiles were developed from collaborations between Nuno and small-medium sized manufacturers (SMMs). However, Reiko Sudo also initiated projects with companies from diverse industrial sectors, including companies producing beer, tyres, and food as described in vignettes 'industry scanning and serious play – jellyfish fabric', 'metallic plating of textiles', and 'collaboration in the creation of eco-textiles'.

Even the titles of the books are poetic; they refer to the tactile qualities of the fabric as well as the emotional response they induce (R. Sudo, personal communication, May 2006). *Shimi Jimi* (evocative, engaging, poignant), *Fuwa Fuwa* (fluffy, flimsy), *Zawa Zawa* (noisy, agitated), *Boro Boro* (ragged, fragile), *Kira Kira* (glittering, shiny – which includes descriptions of the innovative 'melt-off technique' that characterizes the integration of metals in textiles), and *Suke Suke* (sheer, transparent). I am convinced that the primary motivation for the collaborations to create textiles and books was not financial. The conscription device can be described as a desire and commitment to share this creative journey with a wider community of artists. Textile students from art schools all over the world have found inspiration in these books.

The scale and scope of an innovation partnership makes a huge impact on what can be achieved. Specific questions need to be asked about the purpose of a partnership, *who* is involved, *why* their contribution is vital, and *how* the individuals and organizations will work together. This can be illustrated by reference to Table 10.1.

Partnership involves cooperation based upon a shared vision and mutual trust. Partnership does not just happen because people meet occasionally. It requires each organization to commit time and effort to create something new together. It involves abandoning old routines and working with partners to devise new practices, which are orientated towards achieving a shared and clearly articulated goal, embodied in the emerging object.

Policy makers have mistakenly assumed that 'access to finance' can unlock the innovation potential in partnerships. Many instances of policy borrowing, inspired by successful innovation in Silicon Valley, have assumed that the most important enabling factor is access to finance. For instance, the HTC policy makers designed all sorts of ways to bring diverse organizations together around a policy goal with adequate resources. However, money did not seem to be very effective as a conscription device in multi-agency partnerships as illustrated in vignettes 'sustaining coherent development with short-term projects' and 'conflict between the roles of animateur and auditor' from the HTCs, which describe instances where money acted as a barrier to innovation. Money proved to be moderately successful in aligning the interests of partners, and it is an essential enabling factor for innovation, but it did not create the emotional commitment crucial to sustained cooperation.

Table 10.1 Differences between complex and simple partnerships

	Complex partnership	*Simple partnership*
Who?	HTCs included more than ten complex public sector organizations and numerous private sector partners (vignette 7.1.2)	Nuno textile innovation project involves a lead designer working with a project co-ordinator in a single small- and medium sized manufacturing company (vignette 5.2.1)
Why?	Policy makers were reluctant to exclude any organizations from the partnership. Although their contribution was unclear, it was assumed that leaving an organization out would reduce the impact of the partnership (vignettes 7.1.1 and 7.1.2)	A clear project goal and communication process was established at the outset. The scale and scope of the project encouraged 'trial-and-error' experiments within this framework (vignettes 5.2.1 and 6.4.2)
How?	It was almost impossible at the outset to predict how the long-term policy goal could be achieved. People and organizations with no prior history of collaboration learned how to work together over several years. Ambiguity and uncertainty about the best way to achieve the overall policy goal made the process of partnership very challenging (vignette 7.1.1)	Two companies collaborated in a project managed by the interaction of two individuals. They used samples and prototypes as boundary objects to help them work towards a clear, shared 'object' over a sustained period of time (vignette 5.2.1)

The impact of money is determined by contextual factors that influence the outcomes of improved 'access to finance'. Public funding cannot act in the same way as investment from 'venture capitalists' to stimulate innovation. Although money from the private sector can catalyse change, public funding can have a negative impact on innovation by creating a powerful bureaucracy that acts as a barrier to change. This is illustrated by reference to the unique financial climate in Silicon Valley described in vignettes 'the role of venture capitalists in Silicon Valley' and 'the unique nurturing environment in Silicon Valley'.

Attitudes to financial risk and the acceptance of the possibility of failure are vital ingredients for innovation. The bureaucracy associated with auditing the prudent use of public money is at odds with the flexibility needed to foster innovation. This limits the effectiveness of money as a 'conscription device'. There seems to be no substitute for a strong emotional commitment to the object of activity. In summary, Table 10.2 sets out some key points from my analysis of case study material.

Table 10.2 Factors critical to successful innovation partnerships

Key factor	*Practical example*
The structures and processes adopted in multi-agency partnerships have a strong influence on how well people can collaborate	A complex implementation chain and three levels of decision-making in the HTCs impeded collaboration between policy makers and practitioners. The tendency for one-way communication made it difficult for practice to inform policy (vignette 7.1.2)
In the desire to emulate success stories, there is a risk that a recipe successful in one context is transplanted or partially copied. Often this policy borrowing misses some key contextual attributes	Attempts to replicate the innovation success in Silicon Valley in the HTCs omitted key factors. The impact of significant contextual differences in the financial services in the US was not fully appreciated. Attitudes to the risk of failure were entirely different in the UK and the US (vignette 7.1.1)
The need to manage the accountability of the partners needs to be balanced with the powerful inertia caused by bureaucracy	Public funding attempted to fill the space occupied by venture capitalists in the US. In attempting to show that public money was spent wisely, a bureaucracy was created where decision-making was so slow and cumbersome that private sector partners became very frustrated. A mechanism that was intended to act as a catalyst for innovation became a barrier (vignette 6.4.1)
Attitudes to risk have an impact, and tolerance of risk seems to be a critical success factor for innovation	Silicon Valley accepts business failure as an inevitable consequence of innovation. A high proportion of start-up businesses expected to fail. If the technology is 'promising', the entrepreneur does not suffer from the stigma of 'business failure'. Larger more stable corporations tend to acquire failed companies for the potential of the technology. Several failed business ventures are perceived as strength and not as a weakness because they are understood to be learning experiences (vignette 7.2.3)

(*Continued*)

Table 10.2 Continued

Key factor	*Practical example*
Some of the most successful partnerships in the case studies were effective because they were small in scale and time-limited experiments that allow the partners freedom to manoeuvre and respond to changing circumstances	Initially, an ill-conceived notion of business planning constrained practitioners. Eventually, discrete projects within the complex HTC programmes of work allowed experimentation to take place. The small-scale project partnership created new organizational practices and built trust between organizations, so they could work together on more ambitious challenges. Jumping straight into a complex, long-term innovation is bound to be difficult (vignette 7.1.2). By contrast, Nuno projects exemplified successful small-scale project partnerships (vignette 6.4.2)
Because relationships change over time, it is valuable to periodically review the divisions of labour and rules and membership of innovation partnership projects	Even in small Nuno partnerships, relationships shift and new roles evolve over time. A review and evaluation of the roles and responsibilities will allow a partnership to be more reflective and responsive to the context in which people are working (vignettes 6.4.2 and 7.3.1)
Keep tracking the object over time. The object of activity cannot be taken for granted, it is both changeable and elusive	In the US, the SMUD organization developed over several decades to take on a mediation role between community groups and commercial operators. In the 1970s, the 'object' was to protect the environment by decommissioning a nuclear plant, whereas in the last few years, the emphasis has been on reducing energy consumption by ecologically sustainable technologies (vignette 7.2.2)

This chapter highlights the characteristics of partnerships that emerged from my detailed case studies, but I have also drawn on the work of Ralph Stacey (1993, 1996, 2002, 2003) who applied concepts from complex adaptive systems (CASs) theory to studies of organizational management. Stacey devised the Difference Matrix as a frame to help people think about the emerging patterns of a group's behaviour. It was designed as a tool to help people to plan an activity, observe an interaction, or intervene to increase a group's capacity for adaptation. It is also possible to use it to understand and encourage emerging systemic change in organizations.

> Difference is the source of creative change and learning. In the same way that a difference in height releases the power of gravity when a river flows to the sea, differences in a group provide the potential for movement and change.
>
> (Stacey, 2006)

This emphasis on 'difference' is conceptually close to the descriptions of clashing organizational perspectives in ELT. In both instances, differences create tensions that can change entrenched behaviour. As with Engeström's notion of the CZPD, if the proposed change is too great, it will be resisted because it is too challenging. 'Difference' also relates to the gap between old and new practices, so when this gap is neither too great nor too small, there is the best chance of organizational learning.

The following diagrams are adapted from Stacey's (2006) Difference Matrix. When there is no difference among the members of a group, everyone may be quite comfortable, but there is no motivation to get the group moving. Stacey uses a simple diagrammatic format to describe variations in the quality and intensity of interaction (Figure 10.1).

As I have argued throughout this book, partnership involves changes to organizational practices so that partners develop new ways of working. This suggests that interaction in a group provides the opportunity for change. If there is little or no interaction, then organizations or groups get stuck in their different perspectives, and no system-wide emergent patterns are possible. If there is too much interaction, the group may shift its focus from one point to another without spending enough time or energy to resolve any specific issue.

Figure 10.2 shows that partnerships do not stay the same but evolve over time, which means it is likely that they move quadrants during the course of their history. In this sense, the vignettes are like 'snapshots of a process' and describe characteristics evident at a particular phase in the development of a partnership. The lower left quadrant may appear to be unproductive. However, I believe it corresponds to a moment of provisional stability or a breathing space in the midst of turbulent change. Rest and recovery is an essential part of the 'process of change' and may compensate for the turbulent activity described in the upper right quadrant.

In other words, the quality and intensity of interaction changes over time and so does the degree of difference between the partners. Stacey draws attention to some negative consequences when differences are too great, and the group may be unable to agree.

Figure 10.3 shows that if everything ends up in the top left quartile, then every aspect of the organizational partnership is challenging. If most things fall within the bottom left quartile, this indicates a great deal of effort is needed to improve communication between the organizations within the partnership. If everything ends up in the bottom right quartile, it is unlikely that a really exciting innovation will happen as a result of the organizations working together. The

Positive scenario	High difference	Low difference
High interaction	**1** Learning Growth Self-organization	**2** Celebration Reinforcement Energy
Low interaction	**3** Reflection Safety	**4** Comfort Belonging Rest and recovery

Figure 10.1 Potentially positive effects of difference between organizations.

Positive scenario	High difference	Low difference
High interaction	**1** The central technology belt HTC chose to register as a limited company when partners established trust and confidence (vignettes 5.2.4 and 8.2.3)	**2** The ICT cluster of companies chose to act as a partnership, using complementary skills to win bigger contracts (vignette 6.3.4)
Low interaction	**3** HTC3 steering group meets less often with policy makers after a turbulent disagreement (vignettes 5.3.5 and 8.2.1)	**4** Photonics cluster in Japan decreases the intensity of meetings following a successful bid for government grant (vignette 6.3.2)

Figure 10.2 Positive examples from the vignettes.

Negative scenario	High difference	Low difference
High interaction	**1** Stress Conflict Exhaustion	**2** Factions Wasted energy Low productivity
Low interaction	**3** Isolation Misunderstanding Frustration	**4** Boredom Stagnation

Figure 10.3 Potentially negative effects of difference between organizations.

challenge and risk can be rebalanced and fine-tuned. Figure 10.4 illustrates how this can be managed by reference to examples from the vignettes.

Stacey devised a second version of the Difference Matrix as a tool that could be used to support groups through system-wide change and to help them to build new systems. Figure 10.5 shows how a facilitator might intervene to influence the trajectory of change. It also shows how relationships between organizations can be changed when a facilitator intervenes to move people into a new quadrant in the Difference Matrix (Stacey, 2006). Again, there is a similarity between this approach and the way in which researchers intervene at the Change Laboratory in Helsinki to push change in a particular direction.

Figure 10.5 shows how a skilled facilitator can help an innovation partnership to function effectively by intervening when disagreements are heated and helping people move away from the most stressful and risky quadrants. Conversely, a facilitator can energize a stalled partnership by focusing on new challenges and moving people into a more interactive quadrant. Figure 10.6 shows how a facilitator might intervene in different circumstances to shape a positive outcome by reference to some of the vignettes.

Negative scenario	High difference	Low difference
High interaction	**1** Tensions between universities and policy makers responsible for funds due to difference in practical experience of innovation (vignettes 7.1.1 and 7.1.2)	**2** Automotive cluster impeded by focus on minutiae of past history and power struggles between factions (contrasts with vignette 7.4.3)
Low interaction	**3** Disagreements about funding 'high-level' training due to divisions of labour and differences in rules applied to universities and LSCs (vignettes 5.2.4, 8,2,2 and 8.2.4)	**4** Partners rest on their laurels, fail to spot new opportunities and fail to meet to take radical action (vignette 6.3.2)

Figure 10.4 Negative examples from the vignettes.

	High difference	Low difference
High interaction	**1** To move to low difference Share a belief or value	**2** To move to low interaction Explain yourself or leave
Low interaction	**3** To move to high interaction Listen more and Ask a question	**4** To move to high difference Amplify little differences

Figure 10.5 The role of a facilitator in changing relationships between organizations.

So, the facilitator may be continually re-balancing the capacity for innovation with the risk of the partnership breaking apart under the pressure of disagreements. From close analysis of the vignettes, it is clear that a cycle of creative destruction, of stability, and of instability characterizes innovation partnerships. The dialectic between stability and instability is at the root of the tension between the conditions associated with innovation and partnership encapsulated in the paradox of 'creative destruction'. Partnership and innovation seem to require different conditions and relationships to flourish. The two are mutually contradictory.

Partnership involves bringing diverse organizational cultures together with the intention of aligning their commitment to a more expansive object. Radical innovation is at odds with a harmonious partnership because contradiction and creative tension between old and new organizational practices is the essential

	High difference	Low difference
High interaction	**1** To move to low difference *Yes, I can see your point, and we agree on this issue so let's do it. (vignette 6.4.4)*	**2** To move to low interaction *Ok, we have solved this. Let's meet less frequently. (vignette 6.3.2)*
Low interaction	**3** To move to high interaction *Why have we always done it this way? How about trying a new approach? (vignette 5.3.3)*	**4** To move to high difference *Well – we don't have the same skill sets in ICT – so let's focus on adding value by joining forces. (vignette 6.3.4)*

Figure 10.6 Examples of interventions by a facilitator.

ingredient that drives innovation, but it destabilizes a partnership. However, Stacey's model does not embrace this dilemma since it fails to emphasize the tension between partnership and innovation. I have therefore adapted Stacey's Difference Matrix to draw attention to the contradictions inherent in each quadrant as shown in Figure 10.7.

Figure 10.7 suggests that a reliable choice in the early stages of a complex innovation programme would be for quadrant 2 because this balances a fairly good potential for innovation with a low chance of disagreement between the partners. At a later stage in the evolution of the partnership, when partners have more trust and confidence in one another, it would then be viable to move to quadrant 1.

Nevertheless, when applying this method of analysis to complex interactions between organizations, the generic use of the term 'difference' is insufficiently sensitive. Difference needs to be defined with respect to specific facets in the partnership. For instance, if the difference in work cultures is small, then the partnership is likely to be relatively harmonious, but this may not stimulate innovation. Some concrete examples can illustrate the tensions between innovation potential and harmony between the partners. Figure 10.8 refers to vignettes that highlight the risk associated with attempting radical, complex change too quickly, with little prior experience and conversely missing opportunities by staying within a 'comfort zone'.

Diagnostic tools can help actors and organizations anticipate likely difficulties so that when these emerge, they are prepared to respond. This does not suggest there is any substitute for investing time and effort in developing relationships of trust between partners. I am not suggesting there are any template solutions that

Cultural practices	High difference	Low difference
High interaction	**1** Maximum potential for innovation Some **risk** to developing a harmonious partnership	**2** Modest innovation likely Maximum chance of **harmony** within the partnership
Low interaction	**3** Moderate potential for innovation Maximum chance of conflict within the partnership	**4** Innovation unlikely Little communication between partners

Figure 10.7 Tensions between conditions for innovation and partnership.

can be transplanted from one context to another. It is important to recognize that organizations working together in complex innovation programmes can expect to find the process challenging and difficult. However, it offers some comfort and reassurance to know that difficulties can be overcome, and it is useful to be able to use tools and techniques that have been effective in particular contexts. Table 10.3 is adapted from a diagram devised by Blacker *et al.* (2000: p. 290) and proposes a series of questions. These questions can be used to tease out some of the complex and contradictory interactions between factors that foster innovation and those that support the development of partnerships between different types of organizations.

Clearly, multi-agency partnerships vary considerably in their composition. However, for the most complicated partnership structures, this kind of questioning can yield valuable diagnostic information and aid reflection and decision-making about next steps to improve collaboration. The case studies highlighted the interplay of an array of factors critical to successful 'innovation partnerships'.

Table 10.4 proposes a fine-grained analysis of the 'extent of difference' between organizations that can enable a facilitator to intervene to influence specific outcomes. The vignettes associated with the HTC (e.g. 'collaborative projects to develop electronic tools') can be contrasted with the vignette 'employee creativity – the object of Google activity', which describes how Google approaches innovation by virtue of self-organizing project teams. These vignettes also highlight the impact of the scope and scale of innovation on how difficult it is for a large number of different organizations to work together. Google is very selective and does not form partnerships but instead acquires companies with something to offer. Public funding is not used at all because Google has immense financial resources and can operate in a very flexible way with highly motivated and creative employees. Google offers unique benefits to key employees who are treated as valuable researchers and enabled to contribute to the overall success of the company. This is very unusual in the commercial world, but it is a

Cultural practices	High difference	Low difference
High interaction	**1** Ambitious and complex innovation potential in HTCs (vignettes 5.2.4 and 8.2.4)	**2** Nuno textile designer led the collaboration with a tyre manufacturer to develop stainless steel textiles (vignette 6.4.3)
Low interaction	**3** SMUD funded research into new solar roof tiles for zero energy homes (vignette 7.4.10)	**4** Collaborative research into ICT skills gaps and local training needs (vignette 6.3.4)

Figure 10.8 Contradictory factors driving innovation and partnership.

source of competitive advantage that I believe will have a huge impact on future business culture.

In contrast to the rapid acquisition of companies by Google, the HTCs created a cumbersome and bureaucratic 'innovation partnership' with long-term plan and huge ambition. The achievements of the HTCs under very difficult circumstances should not be underestimated. Figure 10.9 illustrates the value of mini-expansive

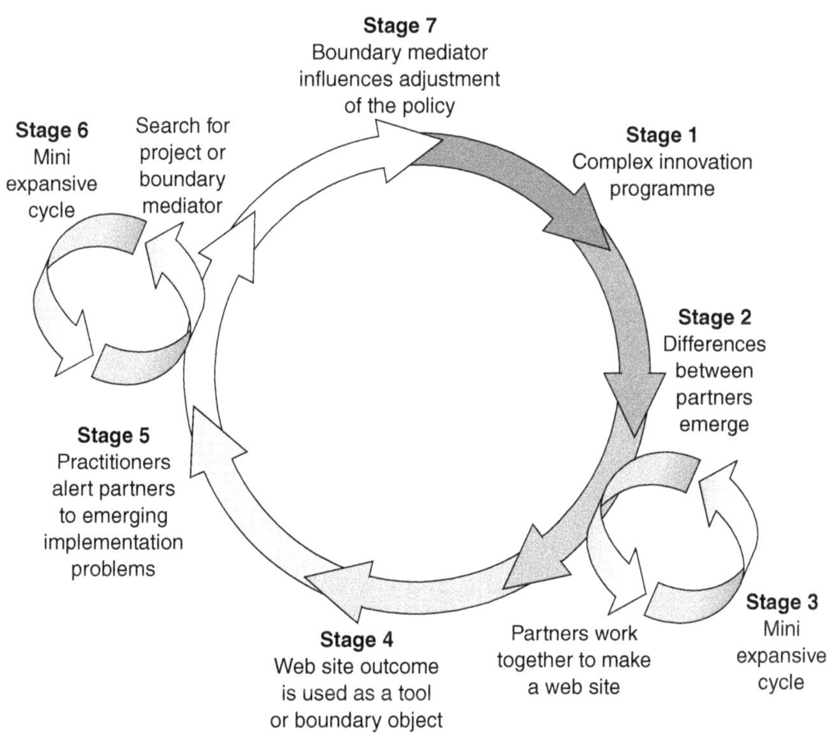

Figure 10.9 The cumulative impact of mini-cycles of expansive learning.

Table 10.3 Analysis of expansive learning in complex innovation partnerships

Analysis of complex innovation programmes	*Complex activity networks*	*Power and agency*	*Developmental history*	*Contradictions*	*Expansive cycles*
Who are learning?	How many partner organizations are engaged?	How much do the various organizational **cultures** differ?	How much experience of working together exists in the partnership?	Are there tensions between relatively weak and powerful organizations?	Are there small successes, projects, or outcomes?
	What is the **division of labour**?	Whose voices are dominant?		Are there **contradictory rules** and/or divisions of labour between levels in implementation **hierarchy**?	Have these **outcomes** been used as **tools** to build trust and bigger expansive cycles?
	How **complex** is the innovation programme?	What are the power dynamics?			
Why do they learn?	What is the espoused object?	Does the espoused object prevail as a dominant discourse?	Have **competing discourses** emerged over time?	Are competing objects affected by **political sensitivities**?	What motivation drives collaboration between levels?
	Are there **competing objects** driven by different partners?		Is there a change in the policy discourse?	Are there different perceptions of **risk**?	
What do they learn?	How do differences of **perspectives** drive innovation?	What strategies help shared solutions to problems to emerge?	Are people able to take more **risks** and to trust the other partners?	Do conflicts between different perspectives create new ways of doing things?	Are people confident that they can do more together than by working independently?

	Do these differences make **relationships** between partners challenging?			How are disagreements resolved? What is the **risk** of failure?	Is a **co-operative partnership** possible?
How do they learn?	Are **boundary objects** created and used to connect activity systems?	How is a need for **harmony** in the partnership balanced with **argumentation** that drives learning?	How are **boundary objects** developed and enriched to help partners move across activity systems?	How do **boundary mediators** support learning between levels separated by agency?	How is **evaluation** used to support expansive learning? How **sustainable** are new practices?

Source: Blackler *et al*. (2000: p. 290).

Notes

Complex innovation programmes are likely to be driven by knowledge transfer policy. Such policies can be enacted in national regional, and local contexts. 'Triple helix' interactions between industry, academia, and government can be visualized as interacting activity networks in which an activity system is the smallest unit of analysis.

Table 10.4 Examples of mediation by facilitators

Dimension of difference	*Role of facilitator*	*Mediation techniques*
1 Experience of innovation	Intervene to share the benefits of experience. Aim to reduce the of gap between old and new practices (the CZPD)	Encourage experienced partners to describe practices in past projects. Use as boundary objects (or tools) to connect the activity systems of the partners
2 Risk associated with the endeavour	Inject a sense of realism about the balance between costs and benefits of an ambitious programme	Focus on experimental projects that allow rule breaking and rule making. New rules can act as tools to shape new attitudes and practices
3 Political sensitivity and visibility	Focus on the experimental nature of the work and intervene to build trust between the partners, by small steps (mini-expansive cycles)	Encourage the partners to create a boundary object, such as a Website or a project that is owned and used by everyone
4 Power dynamics (dominant organizations)	Empower the weaker partners and enable their voice to be heard. Support the expansion of the object by mediation or creation of boundary objects	Persuade powerful strategic actors to mediate as champions on behalf of practitioners. Analysis involves tracing conflicting objects and outcomes at different levels in the hierarchy
5 Complexity (a long implementation chain)	Raise awareness of the value of the feedback loops between practice and policy	Use projects as tools to demonstrate the communication benefits and 'efficiency gains' of reducing the implementation chain
6 Evaluation	Promote the use of evaluation as a tool for learning and reflection. Avoid the victory narrative	Treat evaluation as a continuous process – a research project that enhances engagement and strengthens a partnership

cycles that arose from 'problems', which created outcomes that were used as tools in a larger expansive cycle of innovation.

This model recognizes the need for practice to inform new theory. Tools are created as outcomes of activity and can be used to help the partnership adapt to cope with new circumstances. There is a need for a continuous re-balancing of tensions, which creates a highly stressful environment in which actors and organizations are expected to take risks and do their best creative work. There are no easy prescriptions for making all this easier.

As I have argued throughout this book, innovation is a process of 'creative destruction', which depends on robust exchanges of ideas between the various parties as old and new ways of doing things come into conflict. The contradictory requirements to simultaneously foster both 'partnership and innovation' are at the root of the difficulties in delivering innovation and technology transfer by means of complex triple helix policy programmes

11 Conclusions

11.1 Risks and limitations of the research

There is a risk that my analysis focuses on aspects of linear rationality and coherent planning. Throughout this book, I have emphasized that partnerships are often formed without any clear idea of the contribution each organization can make. I have also argued that innovation is essentially unpredictable and experimental, and it is therefore important to allow flexibility in the way in which multi-agency partnerships address the challenge.

By highlighting some insights about what works well and what tends to be problematic, it is possible to identify effective strategies and potential hazards, and this is relevant to the enactment of similar policy initiatives elsewhere. My international comparisons of 'innovation by partnerships' emphasized the impact of the scale and scope of the 'object of activity' and the significance of different contexts. These cases showed that 'complexity' is a huge obstacle to innovation partnerships. Google is hugely successful and almost synonymous with the term 'innovation'. The method of encouraging self-organizing teams to pursue ideas, which are peer reviewed via their ORKUT 'social networking software' (also developed by colleagues), is really easy to facilitate. It simply requires managers to employ talented and creative people and stand back. The Nuno Corporation empowers designers to develop visual ideas, which may be selected for commercial development by the head designer. Collaboration with a textile manufacturing company or with a company from a different industrial sector is undertaken only if there is a clear aim. Boundary objects are used to bridge the gaps between imagined possibilities and practical outcomes, and relationships between the companies are carefully managed.

The UK case studies involved the most complex partnerships, and fortunately, I was able to meet the key actors and study detailed documentation in the UK. There is a risk that this research could be interpreted as a criticism of the people and institutions responsible for enacting the regional and national innovation policy in the UK. Consequently, it is important to stress that the region coped admirably with an almost impossible challenge. No doubt, individual interests and organizational positioning may have a bearing on the interpretation of the findings and on perceptions of the 'legitimacy of claims'.

Although there were surprises and outliers in the empirical data, the analysis revealed clear patterns, which shaped insights that are relevant to similar programmes and policy initiatives elsewhere. I began my research in 2003, almost four years after the Rover Task Force had first met to consider the embryonic idea for high-technology corridors (HTCs). In retrospect, the discussions that shaped the policy at this stage were influenced by unique circumstances surrounding the looming closure of the Rover plant in 1999. If it had been possible for me to go back in time and gather data from those debates as they happened, these would make fascinating reading in comparison with the recollections of events as described to me in interviews.

My reconstruction of the *developmental history* of innovation projects relies upon interviews capturing rich contextual data. This included how decisions were reached, and inevitably, recollections are coloured by tendencies to simplify what happened in order to tell a story that 'makes sense'. This does not suggest that people deliberately falsified their memories and interpretation of events. Yet, the freshness and immediacy of how partners worked together to reach decisions fades over time, and all that remains is a story of their cumulative impact. There is a risk that a *post hoc rationalization* of policy leaves out much that is of interest and a valuable resource for learning is lost. Ideally, I would like to have taken five years to track 'changes in practices' and carried out more interviews over a larger sample size, including a fuller investigation of anomalies and outliers such as individuals in organizations who shunned involvement in HTCs, despite being in a position to make a contribution.[1]

Nevertheless, comments made by the respondents in the research illuminate what happened. By triangulating data from multiple sources, I was able to suggest possible reasons for 'implementation problems' and make suggestions about what might have worked more smoothly.

Although expansive learning theory (ELT) was used in studies of organizational change and innovation, there was very little reference to the dynamics of political power and agency (Hasu, 2001; Ludvigsen *et al.*, 2003; Toiviainen, 2003; Tuomi-Grohn, 2003). Recent studies have focused on innovation practices within a single corporation. There were no precedents for studies of innovation by complex multi-agency partnerships and no existing methodology that I could replicate. I suspect that my research may be at the outer edge of what is achievable because of the multi-faceted interactions in HTCs.

As mentioned in Chapter 2, some attempts have been made by researchers in Finland to shape new practices in laboratory environments. In the Change Laboratory, very detailed analyses of interactions were made based upon video evidence. In the laboratory, environment researchers could intervene and deliberately push change in a particular direction. While I could agree that in structured and facilitated conditions it may be possible to depict and influence change, in naturalistic settings, it is very difficult to represent the complex multi-layered interactions within activity networks (Toiviainen, 2003; Tuomi-Grohn, 2003). The object in a complex activity network is inherently problematic because the consistent direction and coherent motivation for change is very difficult. It is even more challenging to identify and track the evolution of the 'object of activity' in a complex multi-agency partnership such as an HTC. Although I was able to gather data for my international case studies by meeting key actors, there were no structures to facilitate reflective practice and to therefore drive the cycle of expansion.

In the Change Laboratory in Finland, researchers are empowered to intervene to push a change in a particular direction. By contrast, my case studies traced the development of the object of activity retrospectively in an attempt to diagnose the effectiveness of collaborative practices and organizational learning. Consequently, it is possible that critics of my research may focus attention on how my methodology differs from that adopted in Scandinavian studies. There is also a risk that critics may focus on 'what my analysis cannot do' and suggestions that my use of ELT 'generates more heat than light'.

One disadvantage of using ELT is that it is inaccessible and obscure to people engaged in facilitating change because of its complex and abstract terminology. Specific meanings of key terms in activity theory are too close to everyday vocabulary in English, and this can be confusing. Much of the subtlety of theoretical concepts devised by the original Russian and Finnish authors is lost when translated into English. For instance, the term object has a specific meaning in activity theory. It seems closer to the meaning of the everyday meaning of the word objective than to the everyday word object. There is also some slippage in the consistency with which terms are used by key authors. The ad hoc introduction of new terms clutters the theoretical vocabulary and tends to obscure, rather than clarify, the meaning of activity in particular contexts.

As mentioned briefly in Chapters 2 and 9, there are risks in using ELT because it is evolving and the terminology can be confusing. Nevertheless, ELT offers an action-orientated framework in which to study the effect of social practices and structures on relationships between individuals in organizational contexts.

11.3 Future development of ELT

My analysis pushed the envelope of ELT to its outer limits. ELT can increase awareness of the developmental trajectory of activity and the potential to mould change to achieve specific outcomes. Although this suggests why change does not happen smoothly and may even reduce the anxiety of change, it does not make change predictable.

However, ELT may need to be radically developed if it is to inform practices rather than simply point to complexity. The challenge is to move from an understanding of 'how things are' towards *imagining* better ways of doing things. The challenge presented by the complexity of networks of organizations, loosely connected in the so-called partnerships, increases the need to develop comprehensive models and tools to help people deal with new work environments.

I anchored my development and adaptation of tools and methodologies in my analysis of practices. By grounding my development of the theoretical landscape of ELT in evidence, it is possible to influence policy makers and strategic actors. This is encapsulated by the enthusiasm of UK policy makers for 'evidence-based practice'. This approach is consistent with Vygotsky's contention that new theory can grow from the interaction between everyday concepts and scientific concepts.

My research revealed that boundary objects play a pivotal role in linking the perspectives of different activity systems. However, they are complex and transitory tools, and so a focused and longitudinal study of the changing nature and form of boundary objects may yield fascinating insights. A study of the evolution of *boundary objects* over several years could illuminate their capacity to change relationships between activity systems (Ludvigsen *et al.*, 2003: p. 308). It could also trace the transformation of boundary objects into elements of activity such as tools and outcomes, as suggested in my research and described in Chapter 5.

In the future, it would be useful to study how to create 'conscription devices' to deliberately enhance 'emotional commitment' to a new partnership. If this was better understood, it could speed up the development of project teams from instrumental alliances towards more robust collaborations leading to sustained cooperation. Conscription devices may reduce anxiety about change and help people to feel positive about new ways of doing things. For instance, Google's 'social networking tool' ORKUT grew from the enthusiasm and personal interests of an employee into an in-house tool for innovation (FastCompany, 2004). Google now provides ORKUT as a free service, but this gives Google access to new communities and valuables of new markets with huge commercial potential.

The dense theoretical vocabulary in ELT is a barrier to communicating some of the most important concepts. The object of activity is the key concept in ELT, but the current theoretical vocabulary is not adequate to the task of describing the relative strength of motivation in terms of the degree of emotional commitment to the object. To resolve this difficulty, I believe that more use could be made of metaphors because they can embrace complexity and contradiction. Metaphors construct visual images that create relationships between different and contradictory phenomena and processes (Fichtner, 1999: p. 315). For instance, the metaphor of *technology pull and push* visualizes the dilemma for 'technology transfer' policy makers concerned with 'demand' and 'supply' of knowledge. The metaphor of *translation* captured the need to connect understanding between two different cultures and to link the knowledge of users and developers of new technology. There were many references in the interview data to *boundary objects* as tools to facilitate *technology translation* which could change the direction of change.

Tracking the object of activity is vital because it changes constantly. Although activity is transient and inherently unstable, there are potentially valuable *breathing spaces* where it is possible to pause for reflection in order to make wise decisions and move forwards. The notions of provisional stability and the 'expanded object of activity' represent moments when it is possible to reconsider the direction and momentum of change (Bonamy *et al.*, 2001; Engeström, 2001).

The attitude of the policy makers at the Regional Development Agency (RDA) towards the HTC programme generated a metaphor of a driver with *one foot on the brake and the other on the accelerator*. It vividly expressed how a confused organizational identity led to consequent frustrations for 'partner organizations'. In some ways, metaphors can enrich descriptions of activity, because they express something of the human emotion that can otherwise get lost in a rather dry theoretical description. For instance, this comment illustrates the communicative power of more vivid language. 'The spirit of the policy was not clear even to those delivering it. There was greater understanding of the policy ideals at grass roots level'.[2] Finally, the creation of the so-called knowledge economy and the rise of global corporations suggests that activity systems, which were previously segregated, will be interlinked and grow larger. Yet, little is known about the consequences of working in increasingly complex work environments mediated by electronic communications.

As new organizational contexts arising from global networking, it would be valuable to orientate the scope of ELT towards these new work contexts. There is potential to explore the relationship between individual and collective consciousness. The so-called freedom of 'knowledge workers' is questionable because they are dependent on new socially embedded *knowledge technologies* that lie beyond their control (Hellström, 2004: p. 641).

Little is understood about the role of *boundary mediators* in electronically mediated work environments. It is uncertain how conscription devices and boundary objects might act to create allegiance to new communities of practice mediated by the Internet. The development of the knowledge economy produces considerable changes in human activity. ELT needs to be extended to explain how activity systems interact in new virtual work contexts. There are new contexts for product development by experts 'collaborating across' different companies in different time zones so that the team works effectively around the clock (on a 24/7 basis).

It is unclear how initiatives such as the HTC policy that encourages regional clustering of companies and regional knowledge economies will interact with trends for globalization. Ironically, an appreciation of global changes in product development processes where 'knowledge workers' collaborate across time zones, and national borders had a positive impact on local clustering behaviour. In at least one HTC, an awareness of global networks caused local companies to see one another not as competitors but as allies in competing for work.

> They might start in Australia, then its passed to someone in India then it comes to the UK – then it goes to United States, and in that circle you're getting people working on that design 24 hours a day – so that what used to take three years to design a new car can now be done in 6 months.[3]

This poses an interesting challenge to the development of organizational applications of ELT. A sustained study of the development of *boundary objects* as tools to bridge the perspectives of remote knowledge workers could be valuable. The economic transformations that flow from a *Knowledge Society* may depend upon how well humans manage to control and utilize technology compared with the moulding of human practices by the dominant work tools. ELT stresses the importance of intention and commitment. My research infers that people are capable of innovating for other purposes than those intended by the powers that be.[4] Complex and rigid long-term plans may be counterproductive to innovation. The approach to innovation adopted by Google may prefigure a new type of corporate culture. The success of the knowledge economy may depend on empowering employee creativity. Successful managers of the future may spend less time developing long-term strategic business plans and more time developing the most valuable assets of a corporation – the people.

Glossary

AfA Agenda for Action, a regional policy document for economic development.

Agency The term is used in two ways; in the context of this study, this means the capacity to act but can also refer to a type of institution or organization such as the Regional Development Agency.

ANT Actor-network theory was developed to explain how technologies have agency and can interact with humans to achieve their effect.

AWM Advantage West Midlands (the Regional Development Agency).

Boundary object A tool to link two or more activity systems, often bridging a cultural divide between different communities of practice.

Co-configuration Expansive learning theory uses this term synonymously to mean co-construction of the object of activity by multiple activity systems.

Conscription device This enrols people into an emotional commitment to a community of practice or activity system.

Collaboration The term used in expansive learning theory to denote organizational learning.

Collective Zone of Proximal Development (CZPD) The gap between old and new organizational practices in complex activity networks (see Zone of Proximal Development).

Consensus Agreement by voluntary basis without financial or other forms of coercion.

COPT Communities of practice theory belongs to the family of sociocultural theory and describes human action within its social context. It describes the induction of a new member, such as an apprentice, into a community of practice as 'legitimate peripheral participation'. The theory has been widely used in understanding small group interactions in the workplace (see Chapter 2).

Developmental transfer This describes expansive learning by large organizations or institutions.

Division of labour This describes how work is divided up according to tasks and roles.

DTI Department of Trade and Industry in UK Government.

ERDF European Regional Development Fund is usually linked to capital projects and expenditure on infrastructure such as land and buildings.

ESF European Social Fund provides revenue to address training needs to meet skill gaps in socially or economically disadvantaged communities. It was part of the Rover Task Force funding package and imposed complex eligibility rules.

Expansive learning This describes an outcome when several activity systems share an emotional commitment and motivation towards a shared goal.

FRESA Regional policy document called the Framework for Regional Employment and Skills Action.

Horizontal learning Describes polycontextuality or learning across cultures.

HTC High-technology corridors are geographically pre-determined sites of interaction between sources of research and commercial exploitation that aim to drive a sub-regional knowledge economy as described in the West Midlands Regional Innovation Strategy.

Knowledge economy The term used to describe a broad family of policies linking economic growth to a skilled workforce and their intellectual resources for wealth creation. The term 'knowledge society' also appears in policy documents, but it is associated with a focus on social networks and has a less instrumental emphasis.

Knowledge transfer The term describes the application of research or new technology in a new commercial context.

Mediation A key concept in expansive learning originating in activity theory that proposed that artefacts can mediate between a subject and an object.

Object The object encapsulates the motivation and trajectory of activity and its orientation towards achieving a shared goal.

Outcomes In expansive learning theory, outcomes are the results of the interaction of six elements of activity within an activity system.

Practices These are routine sets of actions and interactions that can be observed in behaviour and are associated with tacit knowledge.

Polycontextuality A term denoting horizontal learning, often used when experts contribute complementary skills to a demanding problem.

RDA Regional Development Agency (see Agency).

RTF Rover Task Force was set up in 1999 as a macro-level regional forum for public and private sector organizations to work together to address the Rover Automotive Crisis.

Rules These refer to the explicit and tacit rules or norms that structure social relations in an activity system.

Situated practice Within the larger family of sociocultural theory, it emphasizes the situating context of human actions.

Sociocultural theory A category of social theory that has shifted the focus of discussions of 'knowledge management' away from issues of information and technology towards those of human capabilities. Within this family, some theories focus on the dynamics of everyday life and the relationships between people and institutions (*situated practice*), while other theories emphasize the influence of institutions, rules, and norms (*theories of social structure*). Activity theory is an example of a sociocultural theory that includes elements of both within its conceptual framework (see Chapter 2).

Subject A term used in activity theory and used somewhat ambiguously in expansive learning theory to refer to the individual, or small group, or collective body in an activity system.

Technology transfer Similar to knowledge transfer but more specifically concerned with scientific and technological knowledge and skill.

Third mission The third task of universities (in which first and second missions are research and teaching) addresses interaction with industry, commerce, and the wider community.

Tools They act as cultural and historical mediating artefacts in an activity system (such as language).

Vertical learning This refers to learning between activity systems of varying agency and limited by dynamics of politics and power. Vertical relationships between activity systems are uneven and hierarchical. These relationships are different in character to the horizontal relationships between activity systems, which are at the same level.

Zone of Proximal Development (ZPD) A Vygotskian concept describing the optimal distance between an old and a new concept that offer the best opportunity for a child to learn[1] (see CZPD).

Notes

1 Introduction and overview

1 This is the rationale for much recent 'evidence-based policy' in education and healthcare.
2 When 'expansive learning theory' is applied to the study of organizations, the 'multi-agency learning outcomes' are described as collaboration. Empirical data describe both the process, as experienced by key individuals, and specific examples of collaborative practices.
3 The first and second university missions are assumed to be research and teaching. The third mission includes interactive projects with business and the wider community. Higher Education Innovation Fund (HEIF) and Higher Education Reach Out to Business and the Community (HEROBaC) are examples of funding streams, which support the so-called 'third mission' activities.
4 A particular example of this is the innovation activities of Japanese Textile Design Company, which the company would describe in terms of artistic endeavour and creative problem solving.
5 My definitions of vignettes and boundary objects are given in the glossary. They are described in more detail in Chapter 3.

2 The theoretical landscape of knowledge transfer

1 Triple helix theory visualizes interwoven spiral strands of interaction between industry, universities, and government that drive innovation and stimulate a 'knowledge economy'.
2 *The Post-Modern Condition: A Report on Knowledge* (Lyotard, 1984: p. 60).
3 Michel Foucault was a French post-modern philosopher and social critic.
4 Karl Weick describes 'sense making' as a developing set of ideas with explanatory possibilities rather than as a body of knowledge.
5 In activity theory, these routine practices are termed operations.
6 Researchers in the Change Laboratory in Helsinki deliberately intervene to shape change outcomes.
7 An idea conveyed in the political philosophy of G. W. F. Hegel and K. Marx.
8 In the evolution of three generations of activity theory, the work of Leont'ev is the first generation, the work of Vygotsky is the second generation, and expansive learning is the third generation activity theory.
9 Adapted from Engeström (2000) but originally published in 1987 as Figure 2 – The structure of a human activity system on page 78 (Engeström, 1987).
10 Web site reference is www.edu.helsinki.fi/activity/pages/research/transfer.

3 A natural history of my research methodology

1 The circumstances surrounding the threatened closure in 1999 of the Rover plant in Birmingham were described in Chapter 1.
2 This is explained in detail in Chapter 4.
3 P6: Category A – head of innovation at Regional Development Agency.
4 In my UK case studies, I interviewed thirty people working at three levels of HTC policy implementation.
5 Each English region is tasked to produce a local version of this to a standard format determined by the UK Government.
6 In my UK case studies, at the macro-level, I drew on regional policy documents; at the meso-level, I used strategy documents produced by HTC steering groups; and at the micro-level, I took information from project outlines.
7 Vignettes can be used as a methodology for data collection, often in circumstances where sensitive areas of enquiry make it difficult to explore perceptions and beliefs. Responses to vignettes used in this way may elicit something more anonymized and abstract and less threatening than a personal account. This is a specific technique and is discussed at www.soc.surrey.ac.uk/sru/SRU25.html.
8 All the primary documents were stored, coded, and analysed using Atlas ti software. The reference numbers are PD1 onwards.

4 Reconstructing subject perspectives

1 www.centraltechnologybelt.com/ (This is now a commercial operation whereas the Wolverhampton Telford Corridor remains on a different place on a University website. A Google search can locate it easily.)
www.wlv.ac.uk/Default.aspx?page=7140 (The Coventry Solihull Warwickshire Corridor does not have stand alone website although documents are available via Google search in PDF format.) Another safe reference is the website of the Regional Development Agency which holds information on all three HTCs at www.advantagewm.co.uk/working-with-us/high-technology-corridors.aspx
2 Adapted from Figure 7.6 in Toiviainen (2003) and based on the implementation staircase model (Blackler, 1995; Reynolds and Saunders, 1987).
3 P30: Category B – Respondent: manager (representing the Chief Executive Officer) in a public sector organization and member of an HTC2 steering group.
4 P31: Category B – Respondent: head of development in public health sector organization – late involvement with HTC2 steering group and head of regeneration and development of the new super hospital.
5 P9: Category A – Respondent: senior public sector officer responsible for regional innovation contracts.

5 Horizontal expansion of the object by collaboration of experts

1 The Latin root of engineering is the word for 'imagining'.
2 P22: Category C – Respondent: HTC2 project manager.
3 P28: Category B – Respondent: senior manager in public sector business support organization and member of the HTC1 steering group.
4 P12: Category B – Respondent: director of HTC2 project level interview data.
5 The diagram is simplified to show a sample of public and private sector partners, the Regional Development Agency, Higher Education, Learning and Skill Councils, small and medium enterprises, and the National Health Service. Each has its own organizational culture or activity system.
6 Included in a presentation of the key findings from the Research and Innovation Audit for the West Midlands (Brown, 2004b).

7 QinetiQ is a research agency funded by the Ministry of Defence and located in HTC1.
8 P17: Category C – Respondent: manager of a project in HTC 3.
9 P13: Category B – Respondent: director of HTC1 – director of CTB – clarification and project information.
10 P7: Category A – Respondent: pivotal and strategic actor in regional innovation and HTC3.

6 Mediation between activity systems of unequal agency

1 Engeström colonized communities of practice theory (COPT) and describes an activity system as a relatively stable community of practice.
2 P21: Category C – Respondent: manager of an HTC2 project.
3 P12: Category B – Respondent: director of HTC2 project level interview data.
4 P20: Category C – Respondent: manager of an HTC2 project.
5 P20: Category C – Respondent: manager of an HTC2 project.
6 P24: Category C – Respondent: technical manager of an SME and beneficiary of a project in HTC3.
7 P24: Category C – Respondent: technical manager of an SME and beneficiary of a project in HTC3.
8 P24: Category C – Respondent: technical manager of an SME and beneficiary of a project in HTC3.
9 Refer also to the Nuno vignettes for other examples of 'emotional commitment to the object' and 'conscription devices'.
10 P17: Category C – Respondent: manager of an HTC project in HTC3.
11 P23: Category C – Respondent: technical manager of an SME and beneficiary of a project in HTC3.
12 The dotted triangle represents the activity system of the design and marketing team and the striped triangle is the engineers. Beta-testing of the software in a cluster of small engineering companies is represented by the checked triangle, which is the activity system of the HTC project.
13 P23: Category C – Respondent: technical manager of an SME and beneficiary of a project in HTC3.
14 P23: Category C – Respondent: technical manager of an SME and beneficiary of a project in HTC3.
15 Engeström appropriated and developed Star and Griesemer's (1989) concept of *boundary objects.*

7 The impact of complexity and power dynamics on organizational learning

1 Triple helix theory was discussed in detail in Section 2.3.
2 The Rover Task Force was set up on the basis of the first SQW report.
3 P10: Category A – Respondent: mediator and key regional actor for innovation and HTC1.
4 P10: Category A – Respondent: mediator and key regional actor for innovation and HTC1.
5 P27: Category B – Respondent: manager of a science park and member of the HTC1 steering group.
6 P1: head of regeneration at university hospital and CTB board member.
7 P1: head of regeneration at university hospital and CTB board member.
8 P6: Category A – head of innovation at Regional Development Agency.
9 P7: Category A – Respondent: pivotal and strategic actor in regional innovation and HTC3.

10 ERDFs capital developments such as building and site preparation, whereas ESFs training for disadvantaged communities. Each specifies deliverable outcomes such as jobs protected or created some of which are geographically specific.
11 P6: Category A – Respondent: pivotal and strategic actor in regional innovation.
12 P6: Category A – head of innovation at the Regional Development Agency.
13 This was discussed in Chapter 1.
14 From 2003 to 2008.
15 P7: Category A – Respondent: pivotal and strategic actor in regional innovation and HTC3.
16 P25: Category B – Respondent: manager of a science park and member of an HTC steering group.
17 Defence Industry Diversification Agency (DIRA) and QinetiQ.
18 There was an interesting relationship between Coventry University and Jaguar and between Ford and Warwick Manufacturing Group (WMG) given the decision by Ford to close the Jaguar plant in Coventry in September of 2004. Choices were made in terms of strategic priorities. Automotive design expertise and Continuing Professional Development (CPD) facilities at Coventry was not considered to be capable of changing the trend for design to be located at Head Office outside the UK, which was increasingly seen as a centre for assembly.
19 P5: Category B – Respondent: director of HTC2.
20 National centres of private sector research and training for polymers and rubber technologies are located within HTC3, namely the PTA and RAPRA.
21 P15: Category B – Respondent: head of innovation in a rural local authority and actor in HTC1.
22 P2: Category B – Respondent: manager in public sector business support agency and member of the HTC1 steering group.
23 P15: Category B – Respondent: head of innovation in a rural local authority and actor in HTC1.
24 P5: Category B – Respondent: director of HTC2.
25 P14: Category B – Respondent: director of HTC3 project level detail.
26 NOVUS is simply the brand name of an HTC technology transfer project.
27 P16: Category C – Respondent: manager of an HTC1 project.
28 P16: Category C – Respondent: manager of an HTC1 project.
29 P7: Category A – Respondent: pivotal and strategic actor in regional innovation and HTC3.
30 The role of self-organizing teams is related to the concept of emergence in complexity theory.
31 P1: senior figure in public health sector, responsible for regeneration issues, and a member of the HTC1 steering group.
32 P1: senior figure in public health sector, responsible for regeneration issues, and a member of the HTC1 steering group.
33 P1: senior figure in public health sector, responsible for regeneration issues, and a member of the HTC1 steering group.
34 QinetiQ is a newly privatized research facility at Malvern, although it continues to be funded by the UK Government. It is a centre of excellence in liquid crystal displays and microwave technologies and for research into defence medicine. It used to be called the Defence Industry Research Agency (DIRA).
35 P15: Category B – Respondent: head of innovation in a rural local authority and actor in HTC1.
36 University Hospital Birmingham is the second largest hospital trust in the UK.
37 P1: senior figure in public health sector, responsible for regeneration issues, and a member of the HTC1 steering group.
38 Medici is a university project to encourage academics engaged in biotechnology and medical research to spin out research and become entrepreneurial.

39 P1: senior figure in public health sector, responsible for regeneration issues, and a member of the HTC 1 steering group.
40 P1: senior figure in public health sector, responsible for regeneration issues, and a member of the HTC1 steering group.
41 P10: Category A – Respondent: mediator and key regional actor for innovation and HTC1.
42 P21: Category C – HTC project manager and also LSC workforce skills manager.

8 Using evaluation as a tool for learning

1 P34: Category A – Respondent: October 2005.
2 The inference about the significant role of 'communities' draws on both activity theory and communities of practice theory. The mutual influence of these two theories has already been discussed in Chapters 2 and 5.
3 P34: Category A – Respondent: October 2005.
4 P8: Category B – Respondent: key public sector regional innovation officer.
5 P6: Category A – Respondent: pivotal and strategic actor in regional innovation.
6 P34: Category A – Respondent: strongly agree with this analysis when it was presented in 2005.
7 P11: Category B – Respondent: director of HTC3.
8 P18: Category C – Respondent: project manager in HTC3.
9 P8: Category B – Respondent: key public sector regional innovation officer.
10 P7: Category A – Respondent: pivotal and strategic actor in regional innovation and HTC3.
11 P6: Category A – Respondent: pivotal and strategic actor in regional innovation.
12 P6: Category A – Respondent: pivotal and strategic actor in regional innovation.
13 P25: Category B – Respondent: manager of a science park and member of an HTC steering group.
14 P34: Category A – Respondent.
15 P7: Category A – Respondent: pivotal and strategic actor in regional innovation and HTC3.
16 P10: Category A – Respondent: mediator and key regional actor for innovation and HTC1.
17 P5: Category B – Respondent: director of HTC2.
18 P32: Category C – Respondents: managers of projects in HTC2.
19 P25: Category B – director of Warwick Science Park and acted as interim director of HTC2.
20 The source was a conversation on 5 August 2004 with an ex-financial consultant to the RDA.
21 Described in the Rover Task Force Report (SQW, 2001).
22 P34: Category A – Respondent.
23 P15: Category B – Respondent: head of innovation in a rural local authority and actor in HTC1.
24 P15: Category B – Respondent: head of innovation in a rural local authority and actor in HTC1.
25 P25: Category B – Respondent: manager of a science park and member of an HTC steering group.
26 P27: Category B – Respondent: manager of a science park and member of the HTC1 steering group.
27 P28: Category B – Respondent: senior manager in public sector business support organization and member of the HTC1 steering group.
28 P4: Category B – Respondent: director of HTC1.
29 The word 'they' refers to the steering group or board running the HTCs.
30 P8: Category B – Respondent: key public sector regional innovation officer.

31 P20: Category C – Respondent: manager of an HTC2 project.
32 P6: Category A – Respondent: pivotal and strategic actor in regional innovation.
33 P8: Category B – Respondent: key public sector regional innovation officer.
34 P8: Category B – Respondent: key public sector regional innovation officer.
35 P9: Category A – Respondent: senior public sector officer responsible for regional innovation contracts.
36 P9: Category A – Respondent: senior public sector officer responsible for regional innovation contracts.
37 P9: Category A – Respondent: senior public sector officer responsible for regional innovation contracts.
38 P34: Category A – Respondent: October 2005.
39 P27: Category B – Respondent: manager of a science park and member of the HTC1 steering group.
40 The RDA values *outputs* such as 'jobs protected' because they can be achieved quickly and measured. While long-term *outcomes*, such as qualitative changes in practices, are ignored, because they take a long time to achieve.
41 A long implementation chain tends to create a hierarchical structure, with multi-layered 'project management' procedures that disempower practical action and local decision-making.
42 P32: Category C – Respondents: managers of projects in HTC2.
43 ERDFs capital developments such as building and site preparation, whereas ESFs training for disadvantaged communities. Each specifies deliverable outcomes such as jobs protected or created. They also state different eligibility criteria, some of which are geographically specific.
44 P34: Category A – Respondent.
45 P34: Category A – Respondent.
46 One steering group was transformed into a company limited by guarantee.
47 P27: Category B – Respondent: manager of a science park and member of the HTC1 steering group.
48 A recent study depicted communication between hierarchal levels in diagrammatic form (Toiviainen, 2003). These depicted upwards and downwards patterns of learning and transfer between worker and management levels in a Scandinavian manufacturing company, but similar patterns were not evident in my research.

9 Insights from the research findings

1 The EU and DTI policy assumption were discussed in Chapter 2.
2 The concept of 'unfreezing' was originally developed by Professor Lewis Elton, and this was cited in the referenced publication on change.
3 P34: Category A – Respondent.
4 Chapter 2 discussed the conceptual similarities between COPT and ELT in some detail.
5 No one used the term 'crystallized knowledge' to refer to boundary objects.
6 P34: Category A – Respondent: October 2005.
7 P34: Category A – Respondent: October 2005.
8 P34: Category A – Respondent: October 2005 agreed with this conclusion but felt the implications were profound and needed to be backed up with clear irrefutable evidence.
9 The NOVUS programme, Medilink, public interest energy research in California.
10 Examples include the jellyfish fabric, metallic textiles, textile holograms, and environmentally sustainable textiles.
11 When 'expansive learning theory' is applied to the study of organizations, the 'multi-agency learning outcomes' are described as collaboration. Empirical data describe both the process, as experienced by key individuals, and specific examples of collaborative practices.

12 P34: Category A – Respondent.
13 Yet another term is introduced by activity theorists, and 'advanced collaborative practices' are simply practices that can be sustained and become established.
14 P25: Category B – Respondent: manager of a science park and member of an HTC steering group.

10 A toolkit to support organizational learning

1 These books are available at http://www.nuno.com/home.html/. Illustrations of the latest Nuno textiles are also available at this Web site.

11 Conclusions

1 I refer to an anomaly in an interview with an individual at a Defence Research Agency, who was unenthusiastic about the HTC policy.
2 P34: Category A – Respondent: October 2005.
3 P5: Category B – Respondent: director of HTC 2 – director of Coventry Solihull Warwickshire HTC.
4 The success of the Napster, You Tube, and My Space Web sites exemplifies this.

Glossary

1 Vygotsky developed Cultural and Historical Activity Theory in the 1920–1930s and applied this to conceptual development in children. He described how cultural artefacts such as language act as tools to mediate between the subject and object of activity. Engeström developed the idea of a CZPD from this work.

References

1 Introduction and overview

Ball, S. J. (1994). Researching inside the state: issues in the interpretation of elite interviews. In D. Halpin and B. Troyna (eds), *Researching Educational Policy: Ethical and Methodological Issues* (pp. 107–120). London: Falmer.
Beach, D. (2003). A problem of validity in education research. *Qualitative Inquiry*, 9(6), 859–873.
Blackler, F. (1995). Knowledge, work and organizations: an overview and interpretation. *Organizational Studies*, 16, 1021–1046.
Blackler, F. (2000). The rise of objects in the study of organizations. *Organization*, 9(2), 357–358.
Blackler, F., Crump, N., and McDonald, S. (1999). Managing experts and competing through innovation: an activity theoretical analysis. *Organization*, 6(1), 5–31.
Blackler, F., Crump, N., and McDonald, S. (2000). Organizing processes in complex activity networks. *Organization*, 7(2), 277–300.
Engeström, Y. (1999). Innovative learning in work team: analysing cycles of knowledge creation in practice. In Y. Engeström, R. Miettinen, and R. Punamaki (eds), *Perspectives on Activity Theory* (pp. 377–404). Cambridge: Cambridge University Press.
Engeström, Y. (2000). Comment on Blackler *et al.* Activity theory and the social construction of knowledge: a story of four umpires. *Organization*, 7(2), 301–310.
Engeström, Y. (2001). Expansive learning at work: toward an activity theoretical reconceptualisation. *Journal of Education and Work*, 14(1), 133–156.
Engeström, Y. (2004). New forms of learning in co-configuration work. *Journal of Workplace Learning*, 16(1/2), 11–21.
Engeström, Y., Miettinen, R., and Punamaki, R. (1999). *Perspectives on Activity Theory*. Cambridge: Cambridge University Press.
Kerosuo, H. and Engeström, Y. (2003). Boundary crossing and learning in creation of new work practice. *Journal of Workplace Learning*, 15(7/8), 345–351.
Lincoln, Y. S. and Guba, E. G. (2000). Paradigmatic controversies, contradictions, and emerging confluences. In N. K. Denzin and Y. S. Lincoln (eds), *The Handbook of Qualitative Research* (pp. 191–215, specifically p. 181). London: Sage.
Ludvigsen, S. R., Havnes, A., and Lahn, L. C. (2003). Workplace learning across activity systems: a case study of sales engineers. In T. Tuomi-Grohn and Y. Engeström (eds), *Between School and Work: New Perspectives on Transfer and Boundary Crossing* (pp. 291–310). Oxford: Pergamon.
Miles, M. B. and Huberman, M. (1994). *Qualitative Data Analysis: An Expanded Sourcebook* (2nd edn). Thousand Oaks: Sage.

Ozga, J. (1999). *Policy Research in Educational Settings*. Buckingham: Open University Press.

Silverman, D. (2001). *Interpreting Qualitative Data Methods for Analysing Talk, Text and Interaction*. London: Sage.

Toiviainen, H. (2003). *Learning Across Levels: Challenges of Collaboration in a Small-Firm Network*. Helsinki: University of Helsinki.

Yin, R. K. (2003). *Case Study Research Design and Methods* (Vol. 5). London: Sage.

2 The theoretical landscape of knowledge transfer

Bitran, G. R., Gurumurthi, S., and Sam, S. L. (2006). *Emerging Trends in Supply Chain Governance* (Working Report). Massachusetts: MIT Sloan School of Management.

Blackler, F. (1995). Knowledge, work and organizations: an overview and interpretation. *Organizational Studies*, 16, 1021–1046.

Blackler, F., Crump, N., and McDonald, S. (1999). Managing Experts and Competing through Innovation: An Activity Theoretical Analysis. *Organization*, 6(1), 5–31.

Bonamy, J., Charlier, B., and Saunders, M. (2001). 'Bridging tools' for change: evaluating a collaborative learning network. *Journal of Computer Assisted Learning*, 17, 295–305.

Bredgaard, T., Dalsgaard, L., and Larsen, F. (2003). *An Alternative Approach for Studying Public Policy – The Case of Municipal Implementation of Active Labour Market Policy in Denmark*. Available: www.aub.aau.dk/phd/department/02/text/2003/35032003_4.pdf [28 June 2007].

Collins, P., Shukla, S., and Redmiles, D. (2004). *Activity Theory and System Design: A View from the Trenches*. Available: www.ics.uci.edu/~redmiles/publications/J008-CSR02.pdf [28 June 2007].

Colville, I. D., Waterman, R. H., and Weick, K. E. (1999). Organizing and the search for excellence: making sense of the times in theory and practice. *Organization*, 6(1), 129–148.

Cooke, P. and Leydesdorff, L. (2006). Regional development in the knowledge-based economy: the construction of advantage. *Journal of Technology Transfer*, 31(1), 5–15. Also available at: http://users.fmg.uva.nl/lleydesdorff/constructed_advantage/

Davydov, V. (1999). The content and unresolved problems of activity theory. In Y. Engeström, R. Miettinen and R. Punamaki (eds), *Perspectives on Activity Theory* (pp. 39–52). Cambridge: Cambridge University Press.

Engeström, J. (2002). *Organizing Innovation – Innovating Organization: New Product and Service Development at an Internet Consultancy*. Helsinki: Centre for Activity Theory and Developmental Work Research, University of Helsinki.

Engeström, Y. (1999). Innovative learning in work team: analysing cycles of knowledge creation in practice. In Y. Engeström, R. Miettinen, and R. Punamaki (eds), *Perspectives on Activity Theory* (pp. 377–404). Cambridge: Cambridge University Press.

Engeström, Y. (2000). Comment on Blackler *et al.* Activity theory and the social construction of knowledge: a story of four umpires. *Organization*, 7(2), 301–310.

Engeström, Y. (2001). Expansive learning at work: toward an activity theoretical reconceptualisation. *Journal of Education and Work*, 14(1), 133–156.

Engeström, Y. (2003). *Discourse & Object*. T. f. L. University.

Engeström, Y. (2004). New forms of learning in co-configuration work. *Journal of Workplace Learning*, 16(1/2), 11–21.

Engeström, Y., Miettinen, R., and Punamaki, R. (1999). *Perspectives on Activity Theory*. Cambridge: Cambridge University Press.

Etzhowitz, H. and Leydesdorff, L. The endless transition: a 'triple helix' of university industry government relations. Available: http://users.fmg.uva.nl/lleydesdorff/min98/

Fairclough, N. (1999). Global capitalism and critical awareness of language. *Language Awareness*, 3(2), 71–83.

Fichtner, B. (1999). Metaphor and learning activity. In Y. Engeström, R. Miettinen, and R. Punamaki (eds), *Perspectives on Activity Theory* (pp. 314–323). Cambridge: Cambridge University Press.

Foucault, M. (1972). *The Archaeology of Knowledge*. London: Tavistock.

Fox, S. (2000). Communities of practice, Foucault and actor-network theory. *Journal of Management Studies*, 37(6), 853–867.

Fullan, M. (1993). *Change Forces-Probing the Depths of Educational Reform*. London: Falmer.

Fullan, M. (1999). *Change Forces: The Sequel*. London: Falmer.

Gherardi, S. (2000). Practice-based theorizing on learning and knowing in organizations. *Organization*, 7(2), 211–223.

Giddens, A. (1993). *New Rules of Sociological Method*. Stanford: Stanford University Press.

Goddard, J. and Chatterton, P. (1999). *The Response of Higher Education Institutions to Regional Needs* [The Programme on Institutional Management in Higher Education (IMHE) m03/35719]. Paris: OECD.

Halpin, D. and Troyna, B. (1994). *Researching Educational Policy: Ethical and Methodological Issues*. London: Falmer.

Hasu, M. (2001). *Critical Transition from Developers to Users; Activity-Theoretical Studies of Interaction and Learning in the Innovation Process*. Helsinki: University of Helsinki.

Hellström, T. (2004). Innovation as social action. *Organization*, 11(5), 631–649.

Higham, J., Sharp, P., and Yeomans, D. (2002). *Changing the 14–19 School Curriculum in England: Lessons from Successive Reforms. A Summary Report on the Research Project*. Leeds: University of Leeds.

JSBRI. (2005). *Structural Change in Japanese Society and the Dynamism of Small and Medium Enterprises – White Paper* (edited by Small and Medium Enterprises Agency and the Ministry of Economy, Trade and Industry). Tokyo: Japan Small Business Research Institute.

Kerosuo, H. and Engeström, Y. (2003). Boundary crossing and learning in creation of new work practice. *Journal of Workplace Learning*, 15(7/8), 345–351.

Lave, J. (1993). The practice of learning. In S. Chaiklin and J. Lave (eds), *Understanding Practice* (pp. 3–34). Cambridge: Cambridge University Press.

Leydesdorff, L. (2005a). The knowledge-based economy and the triple helix model. In W. Dolfsma and L. Soete (eds), *Reading the Dynamics of a Knowledge Economy*. (pp. 42–76). Cheltenham: Edward Elgar.

Leydesdorff, L. (2005b). The triple helix model and the study of knowledge-based innovation systems. *International Journal of Contemporary Sociology*, 42(1), 12–27.

Leydesdorff, L. (2006). 'While a storm is raging on the open sea': regional development in a knowledge-based economy. *Journal of Technology Transfer*, 31(1), 189–203. Also available at: http://users.fmg.uva.nl/lleydesdorff/storm_at_sea/index.htm [28 June 2007].

Leydesdorff, L., Dolfsma, W., and van der Panne, G. (2006). Measuring the knowledge base of an economy in terms of triple-helix relations among 'technology, organization, and territory'. *Research Policy*, 35(2), 181–199.

Ludvigsen, S. R., Havnes, A., and Lahn, L. C. (2003). Workplace learning across activity systems: a case study of sales engineers. In T. Tuomi-Grohn and Y. Engeström (eds),

Between School and Work: New Perspectives on Transfer and Boundary Crossing (pp. 291–310). Oxford: Pergamon.

Lyotard, J. F. (1984). *The Post-Modern Condition: A Report on Knowledge*. Manchester: Manchester University Press.

Miettinen, R. and Hasu, M. (2004). Articulating user needs in collaborative design: towards an activity-theoretical approach. *Computer Supported Cooperative Work (CSCW)*, 11(1–2), 129–151.

Ozga, J. (1999). *Policy Research in Educational Settings*. Buckingham: Open University Press.

Porter, M. (1998). Clusters and the new economics of competition. *Harvard Business Review*, 76(6), 14–77.

Rebentisch, E. (1994). *New Insights into the International Technology Transfer Process* (Report of Technology and Management Innovation Division). Massachusetts: MIT.

Reckwitz, A. (2002). Toward a theory of social practices: a development in cultural theorising. *European Journal of Social Theory*, 5(2), 243–263.

Reynolds, J. and Saunders, M. (1987). Teacher responses to curriculum policy: beyond the 'delivery' metaphor. In J. Calderhead (ed.), *Exploring Teachers' Thinking* (pp. 195–215). London: Cassell.

Robertson, M., Scarbrough, H., and Swan, J. (2003). Knowledge creation in professional service firms: institutional effects. *Organization Studies*, 24(6), 831–857.

Sapsed, J. and Salter, A. (2004). Postcards from the edge: local communities, global programmes and boundary objects. *Organization*, 25(9), 1515–1534.

Saunders, M. (1995). The integrative principle: higher education and work-based learning in the UK. *European Journal of Education*, 30(2), 203–216.

Stacey, R. D. (1993). *Strategic Management and Organisational Dynamics: The Challenge of Complexity* (4th ed.). Harlow, UK: Pearson Education Ltd.

Stacey, R. D. (1996). *Complexity and Creativity in Organizations*. San Francisco: Berrett-Koehler.

Stacey, R. D. (2002). *The Impossibility of Managing Knowledge* (RSA Lectures). Available: www.thersa.org.uk/acrobat/ralph_stacey270202.pdf [28 June 2007].

Stacey, R. D. (2003). *Complexity and Group Processes: A Radically Social Understanding of Individuals*. Hove: Brunner-Routledge.

Star, S. L. (1989). The structure of ill-structured solutions: boundary objects and heterogeneous distributed problem solving. In M. Huhns (ed.), *Distributed Artificial Intelligence. Volume II* (pp. 37–54). London: Pitman.

Toiviainen, H. (2003). *Learning Across Levels: Challenges of Collaboration in a Small-Firm Network*. Helsinki: University of Helsinki.

Trowler, P., Saunders, M., and Knight, P. (2002). *Change Thinking, Change Practices: A Guide to Change for Heads of Department, Subject Centres and Others Who Work Middle-Out*. Available: www.heacademy.ac.uk/resources.asp?process=full_record&section=generic&id=2 [28 June 2007].

Tuomi-Grohn, T. (2003). Developmental transfer as a goal of internship in practical nursing. In T. Tuomi-Grohn and Y. Engeström (eds), *Between School and Work: New Perspectives on Transfer and Boundary Crossing* (pp. 199–231). Oxford: Pergamon.

Tuomi-Grohn, T. and Engeström, Y. (2003). *Between School and Work: New Perspectives on Transfer and Boundary Crossing*. Oxford: Pergamon.

Viale, R. and Etzkowitz, H. (2005). Third Academic Revolution: Polyvalent Knowledge; the "DNA" of the Triple Helix. *Paper presented at the Triple Helix 5: The Capitalization of Knowledge: Cognitive, Economic, Social and Cultural Aspects*, Turin, Italy.

Weick, K. E. (1995). *Sensemaking in Organizations*. London: Sage.

Weiss, C. (1995). Nothing as practical as good theory: exploring theory-based evaluation for comprehensive community initiatives for children and families. In J. Connell, A. Kubisch, L. Schorr, and C. Weiss (eds), *New Approaches to Evaluating Community Initiatives, Vol. 1: Concepts, Methods, and Contexts* (pp. 1–11). Washington: The Aspen Institute.

Wenger, E. (1998). *Communities of Practice: Learning, Meaning and Identity*. Cambridge: Cambridge University Press.

Wenger, E. (2000). Communities of practice and social learning systems. *Organization*, 7(2), 225–246.

Yanow, D. (2000). Seeing organizational learning: a 'cultural' view. *Organization*, 7(2), 247–268.

3 A natural history of my research methodology

AWM. (2003). *Response from Advantage West Midlands*. Birmingham: Advantage West Midlands.

AWM and AfA. (2002). *The Agenda for Action 2002–4 Advantage West Midlands*. The Regional Development Agency, Advantage West Midlands. Available: http://www.advantagewm.co.uk/work/agenda/ [23 March 2003].

AWM and WMES. (2003). *The West Midlands Economic Strategy*. The Regional Development Agency, Advantage West Midlands. Available: www.advantagewm.co.uk/region/wmes/ [23 March 2003].

Ball, S. J. (1994). Researching inside the state: issues in the interpretation of elite interviews. In D. Halpin and B. Troyna (eds), *Researching Educational Policy: Ethical and Methodological Issues* (pp. 107–120). London: Falmer.

Bauer, M. and Gaskell, G. (2000). *Qualitative Researching with Text, Image and Sounds: A Practical Handbook*. London: Sage.

Beach, D. (2003). A problem of validity in education research. *Qualitative Inquiry*, 9(6), 859–873.

Blackler, F., Crump, N., and McDonald, S. (2000). Organizing processes in complex activity networks. *Organization*, 7(2), 277–300.

Blackmore, J. and Lauder, H. (2005). Researching policy. In C. Lewin and B. Somekh (eds), *Research Methods in the Social Sciences* (pp. 97–104). London: Sage.

Bloor, G. and Dawson, P. (1994). Understanding professional culture in organisational context. *Organization Studies*, 15(2), 275–295.

Byrne, C. (2006). *DTI Technical Textiles Project*. Available: www.smartextiles.co.uk/ [8 February 2006].

Cole, M. (1999). *Cultural Psychology: Some General Principles and a Concrete Example*. Cambridge: Cambridge University Press.

Cole, M. and Engestrom, Y. (1993). A cultural history approach to distributed cognition. In G. Salomon (ed.), *Distributed Cognitions: Psychological and Educational Considerations* (pp. 1–46). Cambridge: Cambridge University Press.

Cole, M. and Wertsch, J. V. (1996). Beyond the individual-social antimony in discussions of Piaget and Vygotsky. *Human Development*, 39, 250–256.

Collins, P., Shukla, S., and Redmiles, D. (2004). *Activity Theory and System Design: A View from the Trenches*. Available: www.ics.uci.edu/~redmiles/publications/J008-CSR02.pdf [28 June 2007].

Davydov, V. (1999). The content and unresolved problems of activity theory. In

Y. Engeström, R. Miettinen, and R. Punamaki (eds), *Perspectives on Activity Theory* (pp. 39–52). Cambridge: Cambridge University Press.

DfES. (2003a). *Developing a National Skills Strategy and Delivery Plan: Progress Report.* London: DfES.

DfES. (2003b). *Developing a National Skills Strategy and Delivery Plan: Underlying Evidence*. London: DfES.

DfES. (2003c). *The Future of Higher Education – White Paper*. London: DfES.

DTI. (2002). *Science into Innovation: Realising the Potential: DTI Manufacturing Strategy*. London: DTI.

DTI. (2003a). *Competing in the Global Economy: The Innovation Challenge-Competitiveness – DTI Innovation Strategy*. London: DTI.

DTI. (2003b). *Meeting the Productivity Challenge 2003 Pre-Budget Report*. London: DTI.

DTI. (2003c). *Minutes of Evidence Taken Before the Science and Technology Committee Contributing to the House of Lords Select Committee Report of Science and Technology and the RDAs*. London: DTI.

Eagar, T. (1986). Technology transfer and cooperative research in Japan. *Welding Research Abroad*, 32(6–7), 2–11.

Engeström, Y. (2001). Expansive learning at work: toward an activity theoretical reconceptualisation. *Journal of Education and Work*, 14(1), 133–156.

Engeström, Y. (2004). New forms of learning in co-configuration work. *Journal of Workplace Learning*, 16(1/2), 11–21.

Engeström, Y., Miettinen, R., and Punamaki, R. (1999). *Perspectives on Activity Theory*. Cambridge: Cambridge University Press.

EU. (2003). *The Role of the Universities in the Europe of Knowledge* (COM(2003) 58 final, Brussels). Commission of the European Communities. Available: http://europa.eu.int/eur-lex/en/com/cnc/2003/com2003_0058en01.pdf [5 February 2003].

Fox, S. (2000). Communities of practice, Foucault and actor-network theory. *Journal of Management Studies*, 37(6), 853–867.

Hasu, M. (2001). *Critical Transition from Developers to Users: Activity-Theoretical Studies of Interaction and Learning in the Innovation Process*. Helsinki: University of Helsinki.

IARPS. (2006). *Industrial & Academic Research Proposition Showcase*. Available: www.idemployee.id.tue.nl/g.w.m.rauterberg/conferences/CD_doNotOpen/ADC/iar.html [8 February 2006].

JETRO. (2004). *Stimulation of Regional Clusters and International Exchange*. Available: www.jetro.go.jp/en/stats/survey/surveys/20040614.pdf#search=%22JST%20%2B%20opto-Science%20%2B%20Japan%22 [2 July 2007].

JSBRI. (2005). *Structural Change in Japanese Society and the Dynamism of Small and Medium Enterprises – White Paper* (edited by Small and Medium Enterprises Agency and the Ministry of Economy, Trade and Industry). Tokyo: Japan Small Business Research Institute.

Kress, G. and Van Leeuwen, T. (1996). *Reading Images: The Grammar of Visual Design*. London: Routledge.

Lave, J. (1993). The practice of learning. In S. L. Chaiklin, J (ed.), *Understanding Practice* (pp. 3–30). Cambridge: Cambridge University Press.

Law, J. (1992). *Notes on the Theory of the Actor Network: Ordering, Strategy and Heterogeneity* (on-line paper). Centre for Science Studies University of Lancaster. Available: www.comp.lancs.ac.uk/sociology/papers/Law-notes-on-ANT.pdf [2 July 2007].

Law, J. (2000). Comment on Suchman, and Gherardi and Nicolini: knowing as displacing. *Organization*, 7(2), 349–354.

Law, J. and Hassard, J. (1999). *Actor Network Theory and After*. Oxford: Blackwell.

Lincoln, Y. S. and Cannella, G. S. (2004). Dangerous discourses: methodological conservatism and governmental regimes of truth. *Qualitative Inquiry*, 10(1), 5–14.

Ludvigsen, S. R., Havnes, A., and Lahn, L. C. (2003). Workplace learning across activity systems: a case study of sales engineers. In T. Tuomi-Grohn and Y. Engeström (eds), *Between School and Work: New Perspectives on Transfer and Boundary Crossing* (pp. 291–310). Oxford: Pergamon.

Martin, M. (2006). *State Red Tape Trips up Green Energy Efforts*. Chronicle Sacramento Bureau. Available: www.sfgate.com/cgi-bin/article.cgi?file=/c/a/2006/09/24/MNGFTLBL0B1.DTL [24 September 2006].

Mason, J. (1996). *Qualitative Researching*. London: Sage.

Miller, G. (1997). Building bridges the possibility of analytic dialogue between ethnography, conversation analysis and Foucault. In D. Silverman (ed.), *Qualitative Research: Theory, Method and Practice* (pp. 24–44). London: Sage.

NEDO. (2006). *Research and Development Collaboration with Universities*. Available: www.nedo.go.jp/english/archives/180425/180425.html

NSF. (2006). *Industrial R&D*. Available: www.nsf.gov/statistics/nsf97324/chp3.htm [2 July 2007].

Prior, L. (1997). Following Foucault's footsteps text and context in qualitative research. In D. Silverman (ed.), *Qualitative Research: Theory, Method and Practice* (pp. 63–79). London: Sage.

San Francisco Chronicle. (2006a). *Hydrogen Powers Energy Hopes: Experts Say it May be the Fuel of the Future*. Available: http://sfgate.com/cgi-bin/article.cgi?file=/chronicle/archive/2001/04/02/MN107106 [27 January 2007].

San Francisco Chronicle. (2006b). *A Lost Opportunity that Worsened Crisis: Utilities and Federal Regulators Shut the Door on Renewable Power in California*. Available: http://sfgate.com/cgi-bin/article.cgi?file=/chronicle/archive/2001/02/12/MN54462 [27 January 2007].

Silverman, D. (2000). *Doing Qualitative Research: A Practical Handbook*. London: Sage.

Troyna, B. (1994). Reforms, research and being reflexive about being reflective. In D. Halpin and B. Troyna (eds), *Researching Educational Policy: Ethical and Methodological Issues* (pp. 1–14). London: Falmer.

United Nations. (2006). *Technology Transfer in Post-War Japan*. Available: www.unu.edu/unupress/unupbooks/uu36je/uu36je05.htm#technology%20transfer%20in%20post%20war%20japan [2 July 2007].

UUK. (2003). *Contributing to the Regions*. London: UK.

UUK and HEFCE. (2001). *The Regional Mission. The regional contribution of higher education. The West Midlands*. Available: www.informaworld.com/smpp/content-content=a713994430-db=all [2 July 2007].

Weick, K. E. (1995). *Sensemaking in Organisations*. London: Sage.

Wenger, E. (1998). *Communities of Practice: Learning, Meaning and Identity*. Cambridge: Cambridge University Press.

Wenger, E. (2000). Communities of practice and social learning systems. *Organization*, 7(2), 225–246.

WMiE. (2003). *West Midlands in Europe Regional Partnership*. Available: www.westmidlandsineurope.org/wmie.html [23 March 2003].

WMiE and ESF. (2003). *The West Midlands European Strategic Framework*. Brussels: West Midlands in Europe. Available: http://www.wmeuropeanstrategy.info/wmeustrategy/downloads/WM%20European%20Strategic%20Framework.doc [23 March 2003].

WMRA. (2003). *The West Midlands Concordat*. Available: http://www.go-wm.gov.uk/concordat [23 March 2003].

Yin, R. K. (1993). *Applications of Case Study Research* (Vol. 34). London: Sage.

4 Reconstructing subject perspectives

Blackler, F. (1995). Knowledge, work and organizations: an overview and interpretation. *Organizational Studies*, 16, 1021–1046.

Blackler, F. (2000). The rise of objects in the study of organizations. *Organization*, 9(2), 357–358.

Blackler, F. and Kennedy, A. (2004). The design and evaluation of a leadership programme for experienced chief executives from the public sector. *Management Learning*, 35(2), 181–203.

Connell, J., Kubisch, A., Schorr, L., and Weiss, C. (1995). *New Approaches to Evaluating Community Initiatives, Vol. 1: Concepts, Methods, and Contexts*. Washington: The Aspen Institute.

Engeström, Y. (1999). Innovative learning in work team: analysing cycles of knowledge creation in practice. In Y. Engeström, R. Miettinen, and R. Punamaki (eds), *Perspectives on Activity Theory* (pp. 377–404). Cambridge: Cambridge University Press.

Fullan, M. (1999). *Change Forces: The Sequel*. London: Falmer Press.

Reynolds, J. and Saunders, M. (1987). Teacher responses to curriculum policy: beyond the 'delivery' metaphor. In J. Calderhead (ed.), *Exploring Teachers' Thinking* (pp. 195–215). London: Cassell.

Toiviainen, H. (2003). *Learning Across Levels: Challenges of Collaboration in a Small-Firm Network*. Helsinki: University of Helsinki.

Tuomi-Grohn, T. (2003). Developmental transfer as a goal of internship in practical nursing. In T. Tuomi-Grohn and Y. Engeström (eds), *Between School and Work: New Perspectives on Transfer and Boundary Crossing* (pp. 199–231). Oxford: Pergamon.

5 Horizontal expansion of the object by collaboration of experts

Aoshima, Y. (1993). *Inter-Project Technology Transfer and the Design of Product Development Organizations*. Cambridge, MA: Massachusetts Institute of Technology.

Berkowitch, J. E. (1996). *Trends in Japanese Textile Technology*. U.S. Office of Technology Policy. Available: www.technology.gov/Reports.htm [2 July 2007].

Brown, D. (2004a). *Confidential Draft Report for AWM – Research and Innovation for the West Midlands* (executive summary). Cambridge: Arthur D. Little.

Brown, D. (2004b). *Research and Innovation for the West Midlands Final Draft March 2004* (quotes from PowerPoint presentation at WMHEA Management Group Meeting, 10 June 2004, Reference 17826). Cambridge: Arthur D Little.

Hasu, M. (2001). *Critical Transition from Developers to Users; Activity-Theoretical Studies of Interaction and Learning in the Innovation Process*. Helsinki: University of Helsinki.

JSBRI. (2005). *Structural Change in Japanese Society and the Dynamism of Small and Medium Enterprises – White Paper* (edited by Small and Medium Enterprises Agency and the Ministry of Economy, Trade and Industry). Tokyo: Japan Small Business Research Institute.

Ludvigsen, S. R., Havnes, A., and Lahn, L. C. (2003). Workplace learning across activity systems: a case study of sales engineers. In T. Tuomi-Grohn and Y. Engeström (eds), *Between School and Work: New Perspectives on Transfer and Boundary Crossing* (pp. 291–310). Oxford: Pergamon.

Millar, L. (ed.). (2005). *2121 the Textile Vision of Reiko Sudo and Nuno*. Epsom, UK: University College for the Creative Arts.

6 Mediation between activity systems of unequal agency

Aoshima, Y. (1993). *Inter-Project Technology Transfer and the Design of Product Development Organizations*. Cambridge, MA: Massachusetts Institute of Technology.

Berkowitch, J. E. (1996). *Trends in Japanese Textile Technology*. U.S. Office of Technology Policy. Available: www.technology.gov/Reports.htm [2 July 2007].

Byrne, C. (2006). *DTI Technical Textiles Project*. Available: www.smartextiles.co.uk/ [8 February 2006].

Christiansen, A. C. (2001). *Technological Change and the Role of Public Policy: An Analytical Framework for Dynamic Efficiency Assessments* (FNI report 4/2001). Available: www.fni.no/publ/FNIreports.html [2 July 2007].

Connell, J., Kubisch, A., Schorr, L., and Weiss, C. (1995). *New Approaches to Evaluating Community Initiatives, Vol. 1: Concepts, Methods, and Contexts*. Washington: The Aspen Institute.

Cooke, P. and Leydesdorff, L. (2006). Regional development in the knowledge-based economy: the construction of advantage. *Journal of Technology Transfer*, 31(1), 5–15.

Eagar, T. (1986). Technology transfer and cooperative research in Japan. *Welding Research Abroad*, 32(6–7), 2–11.

Eagar, T. and Musso, C. (2003). *Role of Technology in Manufacturing Competitiveness* (Working Paper ESD-WP-2003–06). Cambridge, MA: Massachusetts Institute of Technology.

Fine, C. and Raff, D. (2001). Innovation and economic performance in the automobile industry over the long twentieth century. In R. Nelson, B. Steil, and D. Victor (eds), *Innovation and Economic Performance* (pp. 32). Princeton: Princeton University Press.

Hasu, M. (2001). *Critical Transition from Developers to Users; Activity-Theoretical Studies of Interaction and Learning in the Innovation Process*. Helsinki: University of Helsinki.

Hellström, T. (2004). Innovation as social action. *Organization*, 11(5), 631–649.

JETRO. (2004). *Stimulation of Regional Clusters and International Exchange*. Available: www.jetro.go.jp/en/stats/survey/surveys/20040614.pdf#search=%22JST%20%2B%20opto-Science%20%2B%20Japan%22 [2 July 2007].

JSBRI. (2005). *Structural Change in Japanese Society and the Dynamism of Small and Medium Enterprises – White Paper* (edited by Small and Medium Enterprises Agency and the Ministry of Economy, Trade and Industry). Tokyo: Japan Small Business Research Institute.

Leydesdorff, L. (2005). The triple helix model and the study of knowledge-based innovation systems. *International Journal of Contemporary Sociology*, 42(1), 12–27.

Ludvigsen, S. R., Havnes, A., and Lahn, L. C. (2003). Workplace learning across activity systems: a case study of sales engineers. In T. Tuomi-Grohn and Y. Engeström (eds), *Between School and Work: New Perspectives on Transfer and Boundary Crossing* (pp. 291–310). Oxford: Pergamon.

Martin, M. (2006). *State Red Tape Trips up Green Energy Efforts*. Chronicle Sacramento Bureau. Available: www.sfgate.com/cgi-bin/article.cgi?file=/c/a/2006/09/24/MNGFTLBL0B1.DTL [24 September 2006].

Millar, L. (ed.). (2005). *2121 the Textile Vision of Reiko Sudo and Nuno*. Epsom, UK: University College for the Creative Arts.

MIT. (2006). *The International Motor Vehicle Program at MIT*. Available: imvp.mit.edu/frameiv.html#visi [8 February 2006].

NEDO. (2006). *Research and Development Collaboration with Universities*. Available: www.nedo.go.jp/english/archives/180425/180425.html

NSF. (2006). *Industrial R&D*. Available: www.nsf.gov/statistics/nsf97324/chp3.htm [2 July 2007].

Rebentisch, E. (1994). *New Insights into the International Technology Transfer Process* (Report of Technology and Management Innovation Division). Massachusetts: MIT.

Reynolds, J. and Saunders, M. (1987). Teacher responses to curriculum policy: beyond the 'delivery' metaphor. In J. Calderhead (ed.), *Exploring Teachers' Thinking* (pp. 195–215). London: Cassell.

Sako, M. (2000). *Automobile Industry*. Cambridge, MA: Massachusetts Institute of Technology.

San Francisco Chronicle. (2006a). *Hydrogen Powers Energy Hopes: Experts Say it May be the Fuel of the Future*. Available: http://sfgate.com/cgi-bin/article.cgi?file=/chronicle/archive/2001/04/02/MN107106 [27 January 2007].

San Francisco Chronicle. (2006b). *A Lost Opportunity that Worsened Crisis: Utilities and Federal Regulators Shut the Door on Renewable Power in California*. Available: http://sfgate.com/cgi-bin/article.cgi?file=/chronicle/archive/2001/02/12/MN54462 [27 January 2007].

Schrage, M. (2000). *Serious Play*. Boston: Harvard Business School Press.

Tohoku. (2006). *Universities & Research Centres in Japan*. Tohoku University of Art and Design. Available: www.tuad.ac.jp/k/ [8 February 2006].

Toiviainen, H. (2003). *Learning Across Levels: Challenges of Collaboration in a Small-Firm Network*. Helsinki: University of Helsinki.

Tuomi-Grohn, T. (2003). Developmental transfer as a goal of internship in practical nursing. In T. Tuomi-Grohn and Y. Engeström (eds), *Between School and Work: New Perspectives on Transfer and Boundary Crossing* (pp. 199–231). Oxford: Pergamon.

United Nations. (2006). *Technology Transfer in Post-War Japan*. Available: www.unu.edu/unupress/unupbooks/uu36je/uu36je05.htm#technology%20transfer%20in%20post%20war%20japan [2 July 2007].

Wenger, E. (2000). Communities of practice and social learning systems. *Organization*, 7(2), 225–246.

7 The impact of complexity and power dynamics on organizational learning

Berkowitch, J. E. (1996). *Trends in Japanese Textile Technology*. U.S. Office of Technology Policy. Available: www.technology.gov/Reports.htm [2 July 2007].

Bitran, G. R., Gurumurthi, S., and Sam, S. L. (2006). *Emerging Trends in Supply Chain Governance* (Working Report). Massachusetts: MIT Sloan School of Management.

Blackler, F., Crump, N., and McDonald, S. (2000). Organizing processes in complex activity networks. *Organization*, 7(2), 277–300.

Connell, J., Kubisch, A., Schorr, L., and Weiss, C. (1995). *New Approaches to Evaluating Community Initiatives, Vol. 1: Concepts, Methods, and Contexts*. Washington: The Aspen Institute.

Engeström, Y. (1999). Innovative learning in work team: analysing cycles of knowledge creation in practice. In Y. Engeström, R. Miettinen, and R. Punamaki (eds), *Perspectives on Activity Theory* (pp. 377–404). Cambridge: Cambridge University Press.

Engeström, Y. (2004). New forms of learning in co-configuration work. *Journal of Workplace Learning*, 16(1/2), 11–21.

Engeström, Y., Miettinen, R., and Punamaki, R. (1999). *Perspectives on Activity Theory*. Cambridge: Cambridge University Press.

FastCompany. (2004). *Google, Innovation, and the Web*. Available: http://blog.fastcompany.com/archives/2004/03/14/google_innovation_and_the_web.html [2 July 2007].

Goddard, J. and Chatterton, P. (1999). *The Response of Higher Education Institutions to Regional Needs* [The Programme on Institutional Management in Higher Education (IMHE) m03/35719]. Paris: OECD.

Halpin, D. (1994). Practice and prospects in education policy research. In D. Halpin and B. Troyna (eds), *Researching Educational Policy: Ethical and Methodological Issues* (pp. 198–206). London: Falmer.

JETRO. (2004). *Stimulation of Regional Clusters and International Exchange*. Available: www.jetro.go.jp/en/stats/survey/surveys/20040614.pdf#search=%22JST%20%2B%20 opto-Science%20%2B%20Japan%22 [2 July 2007].

Ludvigsen, S. R., Havnes, A., and Lahn, L. C. (2003). Workplace learning across activity systems: a case study of sales engineers. In T. Tuomi-Grohn and Y. Engeström (eds), *Between School and Work: New Perspectives on Transfer and Boundary Crossing* (pp. 291–310). Oxford: Pergamon.

Millar, L. (ed.). (2005). *2121 the Textile Vision of Reiko Sudo and Nuno*. Epsom, UK: University College for the Creative Arts.

Oakland. (2003). *A Framework for Innovation Support Along the A38 Corridor Report Produced for the Advanced Knowledge Alliance*. Cambridge: Oakland Innovation Ltd and Will Rogers Associates.

Porter, M. (1998). Clusters and the new economics of competition. *Harvard Business Review*, 76(6), 14–77.

Schmidt, E. and Varian, H. (2005). *Google: Ten Golden Rules*. Newsweek. Available: www.msnbc.msn.com/id/10296177/site/newsweek/ [2 July 2007].

Schrage, M. (2000). *Serious Play*. Boston: Harvard Business School Press.

SQW. (2001). *Regenerating the West Midlands Region: A Study to Consider Opportunities for High Technology Corridors/Clusters Final Report to the Rover Task Force Stage One* (JC1777). Cambridge: SQW.

Sullivan, D. (2004). *Google Releases Orkut Social Networking Service*. Available: http://searchenginewatch.com/showPage.html?page=3302741 [3 July 2007].

Trowler, P., Saunders, M., and Knight, P. (2002). *Change Thinking, Change Practices: A Guide to Change for Heads of Department, Subject Centres and Others Who Work Middle-Out*. Available: www.heacademy.ac.uk/resources.asp?process=full_record&section=generic&id=262 [2 July 2007].

Tuomi-Grohn, T. (2003). Developmental transfer as a goal of internship in practical nursing. In T. Tuomi-Grohn and Y. Engeström (eds), *Between School and Work: New Perspectives on Transfer and Boundary Crossing* (pp. 199–231). Oxford: Pergamon.

Tuomi-Grohn, T. and Engeström, Y. (2003). *Between School and Work: New Perspectives on Transfer and Boundary Crossing*. Oxford: Pergamon.

Wikepedia. (2006a). *The Dot-Com Bubble.*

Wikepedia. (2006b). *Google.*

8 Using evaluation as a tool for learning

Bredgaard, T., Dalsgaard, L., and Larsen, F. (2003). *An Alternative Approach for Studying Public Policy – The Case of Municipal Implementation of Active Labour Market Policy in Denmark.* Available: www.aub.aau.dk/phd/department02/text/2003/35032003_4.pdf [2 July 2007].

Brown, D. (2004). *Research and Innovation for the West Midlands Final Draft March 2004* (quotes from PowerPoint presentation at WMHEA Management Group Meeting, 10 June 2004, Reference 17826). Cambridge: Arthur D Little.

Christiansen, A. C. (2001). *Technological Change and the Role of Public Policy: An Analytical Framework for Dynamic Efficiency Assessments* (FNI report 4/2001). Lysaker, Norway: The Fridtjof Nansen Institute.

Connell, J., Kubisch, A., Schorr, L., and Weiss, C. (1995). *New Approaches to Evaluating Community Initiatives, Vol. 1: Concepts, Methods, and Contexts.* Washington: The Aspen Institute.

Ecotec and DTI. (2004). *A Practical Guide to Cluster Development.* London: DTI.

Engeström, Y. (1999). Innovative learning in work team: analysing cycles of knowledge creation in practice. In Y. Engeström, R. Miettinen, and R. Punamaki (eds), *Perspectives on Activity Theory* (pp. 377–404). Cambridge: Cambridge University Press.

Ezzy, D. (2001). A simulacrum of workplace community: individualism and engineered culture. *Sociology*, 35(3), 631–650.

Fichtner, B. (1999). Metaphor and learning activity. In Y. Engeström, R. Miettinen, and R. Punamaki (eds), *Perspectives on Activity Theory* (pp. 314–323). Cambridge: Cambridge University Press.

Fine, C. and Raff, D. (2001). Innovation and economic performance in the automobile industry over the long twentieth century. In R. Nelson, B. Steil, and D. Victor (eds), *Innovation and Economic Performance* (p. 32). Princeton: Princeton University Press.

Fullan, M. (1993). *Change Forces-Probing the Depths of Educational Reform.* London: Falmer.

Hellström, T. (2004). Innovation as social action. *Organization*, 11(5), 631–649.

Mickwitz, P. (2003). *Effectiveness Evaluation of Environmental Policy: The Role of Intervention Theories.* Available: www.finnishevaluationsociety.net/sivu.php?artikkeli_id=67 [3 July 2007].

Raisanen, C. and Linde, A. (2004). Technologizing discourse to standardize projects in multi-project organizations: hegemony by consensus? *Organization*, 11(1), 101–121.

Rebentisch, E. (1994). *New Insights into the International Technology Transfer Process* (Report of Technology and Management Innovation Division). Massachusetts: MIT.

SQW. (2001). *Regenerating the West Midlands Region: A Study to Consider Opportunities for High Technology Corridors/Clusters Final Report to the Rover Task Force Stage One.* Cambridge: SQW.

Stame, N. (2003). *Evaluation and the Policy Context: The European Experience.* Australasian Evaluation Society Conference (AES). Available: www.aes.asn.au/publications/ [3 July 2007].

Toiviainen, H. (2003). *Learning Across Levels: Challenges of Collaboration in a Small-Firm Network.* Helsinki: University of Helsinki.

Valovirta, V. (2005). *Learning from the Construction of Programme Theory: Cognitive Mapping in the Evaluation of the Regional Centre Development Programme in Finland.* Available: http://evaluationcanada.ca/distribution/20021010_valovirta_ville.pdf

Vedung, E. (1997). *Public Policy and Program Evaluation.* New Brunswick, NJ: Transaction Publishers.

Vince, R. and Saleem, T. (2004). The impact of caution and blame on organisational learning. *Management Learning*, 35(2), 133–154.

Weiss, C. (1995). Nothing as practical as good theory: exploring theory-based evaluation for comprehensive community initiatives for children and families. In J. Connell, A. Kubisch, L. Schorr, and C. Weiss (eds), *New Approaches to Evaluating Community Initiatives, Vol. 1: Concepts, Methods, and Contexts* (pp. 1–11). Washington: The Aspen Institute.

9 Insights from the research findings

Blackler, F. (1995). Knowledge, work and organizations: an overview and interpretation. *Organizational Studies*, 16, 1021–1046.

Burfitt, A., Gibney, J., MacNeill, S., and Schierenbeck, C. (2003). *A Scoping Study of the Medical Research Community in the West Midlands: A Report on Behalf of Midlands Medici for Advantage West Midlands*: The Centre for Urban and Regional Studies, University of Birmingham.

Colville, I. D., Waterman, R. H., and Weick, K. E. (1999). Organizing and the Search for Excellence: Making Sense of the Times in Theory and Practice. *Organization*, 6(1), 129–148.

Engeström, Y. (1999). Innovative learning in work team: analysing cycles of knowledge creation in practice. In Y. Engeström, R. Miettinen, and R. Punamaki (eds), *Perspectives on Activity Theory* (pp. 377–404). Cambridge: Cambridge University Press.

Fullan, M. (1993). *Change Forces-Probing the Depths of Educational Reform.* London: Falmer.

Fullan, M. (2001). *The New Meaning of Educational Change.* New York: Routledge-Falmer.

Hasu, M. (2001). *Critical Transition from Developers to Users; Activity–Theoretical Studies of Interaction and Learning in the Innovation Process.* Helsinki: University of Helsinki.

Hellström, T. (2004). Innovation as social action. *Organization*, 11(5), 631–649.

Rappert, B. (1999). The uses of relevance: thoughts on a reflexive sociology. *Sociology*, 334(4), 705–723.

Sudo, R. (2006). Response to Helen Brown's First Questionnaire May 2006. Tokyo: Nuno Corporation.

Thompson, M., P. A. (2004). Strengthening Organisational Activity Theory. *Organization*, 11, 5, 579–602.

Trowler, P., Saunders, M., and Knight, P. (2002). *Change Thinking, Change Practices: A Guide to Change for Heads of Department, Subject Centres and Others Who Work Middle-Out.* Available: www.heacademy.ac.uk/resources.asp?process=full_record&section=generic&id-262 [2 July 2007].

Tuomi-Grohn, T. (2003). Developmental transfer as a goal of internship in practical nursing. In T. Tuomi-Grohn and Y. Engeström (eds), *Between School and Work: New Perspectives on Transfer and Boundary Crossing* (pp. 199–231). Oxford: Pergamon.

Weiss, C. (1995). Nothing as practical as good theory: exploring theory-based evaluation for comprehensive community initiatives for children and families. In J. Connell, A. Kubisch,

L. Schorr, and C. Weiss (eds), *New Approaches to Evaluating Community Initiatives, Vol. 1: Concepts, Methods, and Contexts* (pp. 1–11). Washington: The Aspen Institute.

10 A toolkit to support organizational learning

Blackler, F., Crump, N., and McDonald, S. (2000). Organizing processes in complex activity networks. *Organization*, 7(2), 277–300.

Stacey, R. D. (1993). *Strategic Management and Organisational Dynamics: The Challenge of Complexity* (4th ed.). Harlow, UK: Pearson Education Ltd.

Stacey, R. D. (1996). *Complexity and Creativity in Organizations*. San Francisco: Berrett-Koehler.

Stacey, R. D. (2002). *The Impossibility of Managing Knowledge* (RSA Lectures). Available: www.thersa.org.uk/acrobat/ralph_stacey270202.pdf [3 July 2007].

Stacey, R. D. (2003). *Complexity and Group Processes: A Radically Social Understanding of Individuals*. Hove, UK: Brunner-Routledge.

Stacey, R. D. (2006). *The Difference Matrix*. Available: www.siliconyogi.com/andreas/it_professional/sol/complexsystems/TheDifferenceMatrix.html [3 July 2007].

Trowler, P., Saunders, M., and Knight, P. (2002). *Change Thinking, Change Practices: A Guide to Change for Heads of Department, Subject Centres and Others Who Work Middle-Out*. Available: www.heacademy.ac.uk/resources.asp?process=full_record&section=generic&id=262 [2 July 2007].

11 Conclusions

Bonamy, J., Charlier, B., and Saunders, M. (2001). 'Bridging tools' for change: evaluating a collaborative learning network. *Journal of Computer Assisted Learning*, 17, 295–305.

Engeström, Y. (2001). Expansive learning at work: toward an activity theoretical reconceptualisation. *Journal of Education and Work*, 14(1), 133–156.

FastCompany. (2004). *Google, Innovation, and the Web*. Available: http://blog.fastcompany.com/archives/2004/03/14/google_innovation_and_the_web.html [January 2007].

Fichtner, B. (1999). Metaphor and learning activity. In Y. Engeström, R. Miettinen, and R. Punamaki (eds), *Perspectives on Activity Theory* (pp. 314–323). Cambridge: Cambridge University Press.

Hasu, M. (2001). *Critical Transition from Developers to Users; Activity-Theoretical Studies of Interaction and Learning in the Innovation Process*. Helsinki: University of Helsinki.

Hellström, T. (2004). Innovation as social action. *Organization*, 11(5), 631–649.

Ludvigsen, S. R., Havnes, A., and Lahn, L. C. (2003). Workplace learning across activity systems: a case study of sales engineers. In T. Tuomi-Grohn and Y. Engeström (eds), *Between School and Work: New Perspectives on Transfer and Boundary Crossing* (pp. 291–310). Oxford: Pergamon.

Toiviainen, H. (2003). *Learning Across Levels: Challenges of Collaboration in a Small-Firm Network*. Helsinki: University of Helsinki.

Tuomi-Grohn, T. (2003). Developmental transfer as a goal of internship in practical nursing. In T. Tuomi-Grohn and Y. Engeström (eds), *Between School and Work: New Perspectives on Transfer and Boundary Crossing* (pp. 199–231). Oxford: Pergamon.

Index

acquiescence 159
activity theory 20, 190
actor network theory (ANT) 29, 67, 181
agency 51, 181
atlas 36, 185
audit 131, 139, 141, 142, 157, 185

boundary crossing 20, 39, 67, 69
boundary object 20, 26, 39, 56, 82, 110, 154, 175, 181
bureaucracy 5, 142

Cartesian dualism 153
case studies 1–5, 7–11, 16, 23–26, 27–42, 46, 58, 64, 68, 102, 127, 129, 147–156 , 162–170, 175–177, 185
change 9, 17, 41, 110, 161, 176, 184
collaboration 1–5, 8–25, 29–43
communities of practice theory (copt) 16, 67, 76, 154, 181
community 15, 18–20, 24, 29, 42, 44, 47–50, 55, 67–68, 73, 76–82, 87, 99, 115, 132, 149, 162, 181
complex adaptive systems (cas) 16, 165
complexity theory 5, 12, 13, 21–22
conceptual framework 17, 183
context 1–4, 12, 15, 19, 20, 26, 31–32, 42–46, 56, 65, 92, 102, 125, 127, 131, 144, 147, 155, 164–165, 170
contradiction 19, 34, 128, 136, 141–145, 168, 178
cooperation 26, 60, 77, 80–83, 118, 120, 127, 138, 151, 158, 160–163, 178
creative destruction 12, 21, 88, 168, 174
creativity 16, 40, 56, 77, 84, 90, 103, 106, 112, 113, 126, 142, 162, 170, 180
cultural and historical activity theory (CHAT) 16, 54, 190
culture 16, 28, 30, 44–45, 55, 73, 79, 84, 86–89, 113–115, 119, 127–128, 131, 133, 137, 143, 151, 171, 180, 185
CZPD 161, 165, 174, 181, 183, 190

data analysis 5, 31, 36, 41, 43, 154
developmental transfer 20, 158, 181
difference 104, 105, 121, 161, 165–174
discourse 9, 11, 14, 15, 26, 29, 49, 92, 122, 124, 130, 158, 160, 172
division of labour 18–20, 24, 46–50, 54–56, 82, 143, 181
double bind 133

Engeström 4, 9, 17–21, 24, 29, 46, 69, 118, 121–122, 137, 159, 178, 185
ethics 40–42
Europe 13, 71, 85, 94, 99, 126
evaluation 2, 5, 10, 28, 40, 43, 73, 95, 96, 124–146, 151, 165, 173, 174
evidence based practice and policy 125, 184
expansive learning theory 1, 7, 29, 37, 54, 67, 110, 145, 151–153, 156, 158, 161, 176, 181

Foucault 15, 184

Giddens 8, 15, 16

hegemony 141
heuristic framework 3, 4, 46, 55

identity 16, 44, 45, 58, 69, 106, 109, 154, 157, 179
implementation 1–26, 44–52, 59, 65, 68, 69, 80, 92, 94–124, 127–146, 148–157, 161–174, 176, 185, 189
instability 12, 25, 26, 145, 168

interviews 30, 35, 36, 40, 48–50, 131, 140, 176

Japan 1, 13, 14, 27–38, 57–64, 71–76, 86, 89, 102, 103, 108, 147, 149, 167

knowledge economy 1–6, 9, 23, 26, 28–32, 47–48, 94–99, 133, 138, 147, 150, 179, 180, 182
knowledge transfer 1–26, 39, 46, 55, 61, 68, 70, 97, 173, 182

mediating artefact 18
mediation 5, 18, 43, 56, 67–93, 101, 151, 165, 174, 182
methodology 4, 26–43, 44–46, 176, 185
models 8, 10, 15, 25, 33, 34, 57, 66, 147, 177
motivation 68, 79, 83, 106, 108, 110, 155, 157, 162, 166, 172, 176, 178, 182

object 4–5, 8, 18–26, 30, 37, 39, 44–50, 54–66, 77, 80, 82, 92, 99–124, 130, 136, 144–145, 150–153, 156–159, 161–178, 182
ontology 8, 31
organization 15, 16, 21, 25, 28, 39–40, 44, 46, 47, 96, 101, 110, 129, 139, 153, 163–166, 175
outcome 18, 19, 24, 26, 46–53, 56, 57, 66, 70, 77, 79, 81, 82, 92, 129, 154, 160, 167, 171

paradigm 17, 160
policy 1–26, 27–43, 48, 52, 68–93, 125, 132, 147, 159, 174–179, 184
policy borrowing 2, 12, 26, 32–38, 57, 94, 127, 147–148, 163–164
politics 5, 51, 92, 130, 183
positioning 11, 15, 30, 35, 40–50, 110, 152, 175
power 4, 5, 10, 14, 15, 19–22, 28–34, 43, 47, 51, 67–70, 90–92, 94, 96- 124, 128–131, 140–145, 155–159, 161, 165, 168, 172, 174, 176, 179, 183
practice 2, 5, 8, 11, 14–17, 21, 24, 26, 28–30, 35–40, 46, 54, 56–61, 65, 67–76, 94, 101–104, 119, 120, 124, 125–133, 138, 141, 143, 144, 147–158, 161, 164, 174, 177–179, 182
primary data 34, 37, 41
provisional stability 25–26, 118, 166, 178

research claims 31, 41, 175
research design 4, 26, 31, 35, 36, 43
rules 1, 5, 15–17, 19–20, 24, 36, 39, 40, 44–51, 54, 56, 79–82, 97, 130–147, 151–152, 158, 165, 168, 172, 174, 182

sample 35, 56–57, 176, 185
secondary data 35, 37, 41, 49
self-organizing teams 123, 170, 175, 187
Silicon Valley 1, 6, 13, 27, 31, 34–38, 85–88, 96, 98, 104–106, 112, 114, 147–151, 163–164
situated practice 15, 17, 182
social structure 15, 17, 182
subject 4, 17–20, 24, 29–31, 39, 43, 44–53, 54, 81–82, 109, 121, 126, 137, 152–153, 183, 185

technology transfer 1–5, 7, 11, 28, 32, 49–50, 53, 54, 64–65, 69, 70, 73, 75, 94–99, 101, 103, 106. 131, 134, 136, 138, 146–150, 157, 174, 178, 183
tension 16, 19, 21, 31, 41, 52, 75, 135, 139, 141, 144, 152, 158, 168–169
tools 5, 15–20, 24–26, 29, 35, 41, 43, 46–50, 53, 54–59, 66, 79–82, 109–110, 113, 122, 152–155, 161, 169–174, 177–179, 183, 190
translation 12, 37, 105, 108, 178
triple helix 11–13, 34, 67–68, 94, 105, 148, 173, 184

UK 1–3, 11, 25, 27–42, 47, 57–58, 64, 71, 77, 83–88, 93, 102–106, 115, 118, 128–132–146, 147–151, 164, 175–179, 181, 185
USA 1, 12, 27–38, 60, 71–87, 94, 104, 105, 126, 145, 148, 149

validity 41; *see also* methodology
vignette 5, 38, 46, 53, 61–66, 68–69, 92–93, 102, 123, 124. 128–129, 146, 147–157, 161–171, 184

Wenger 17, 28, 67, 120

ZPD 20–26, 183

For Product Safety Concerns and Information please contact our EU
representative GPSR@taylorandfrancis.com
Taylor & Francis Verlag GmbH, Kaufingerstraße 24, 80331 München, Germany

www.ingramcontent.com/pod-product-compliance
Ingram Content Group UK Ltd.
Pitfield, Milton Keynes, MK11 3LW, UK
UKHW021303190926
475701UK00009B/643

* 9 7 8 0 4 1 5 5 4 1 5 8 9 *